Martin Rutzinger, Kati Heinrich, Axel Borsdorf & Johann Stötter (eds.)

permAfrost – Austrian Permafrost Research Initiative

T0135515

Axel Borsdorf, Georg Grabherr & Johann Stötter (eds.)

IGF Forschungsberichte

Band 6

The Series IGF Forschungsberichte / IGF Research Reports aims to present findings of ongoing or recently completed research projects at the IGF which could not be accommodated within the space of an article for a scientific journal. It thus provides a comprehensive picture, freeing readers from the necessity of looking for individual contributions in different journals. The series also documents proceedings of conferences. An interested public will thus find all contributions to a conference theme collected in one volume and will need not to search for them one by one across many different publication series.

Martin Rutzinger, Kati Heinrich, Axel Borsdorf & Johann Stötter (eds.)

permAfrost – Austrian Permafrost Research Initiative

Final Report

Verlag der Österreichischen Akademie der Wissenschaften

Editors
Martin Rutzinger, Kati Heinrich, Axel Borsdorf and Johann Stötter

Coordination: Martin Rutzinger, IGF
Layout: Kati Heinrich, Fides Braun and Valerie Braun, IGF
Cover pictures: IGF
Print: Steigerdruck GmbH, Axams (http://www.steigerdruck.com)

ISBN 978-3-7001-7578-0

Preface

The Austrian Permafrost Research Initiative (permAfrost) is a national project started in 2009 and funded by the research framework program "Alpenforschung" administrated by the Austrian Academy of Sciences. The project has been conducted by five research partners in Austria, which are the Institute of Geography and Regional Sciences of the University of Graz, the Institute of Geology of the University of Innsbruck, the Department of Geography and Geology of the University of Salzburg, the Institute of Geography of the University of Innsbruck. The project was coordinated by the Institute for Interdisciplinary Mountain Research of the Austrian Academy of Sciences. permAfrost is the first Austrian effort to bring together permafrost research on a national scale. Major tasks within the project are the research fields of

- permafrost reformation and degradation
- monitoring climate change impact on rock glacier behavior
- hydrogeological investigation of changed permafrost discharge
- detecting and quantifying area wide permafrost change

The project finalized in 2013 is now looking back to the establishment of a close network of national permafrost activities including a large variety of research activities resulting in several outputs such as scientific papers partly published in index scientific journals, contributions at national and international conferences and workshops, and the elaboration of PhD, MSc and BSc thesis. An overview of the major scientific output from permAfrost is given in the following chapter.

However the finalization of permAforst raises new research questions and challenges. Especially, strategies for establishing and ensuring long-term monitoring and research on alpine permafrost in Austria will be an important task. The editors thank Matthias Monreal for initial project launch and management, Fides Braun for layouting the report, the Austrian Academy of Sciences for financing this project, and especially Dr. Günter Köck for any support given during the time of the project.

Martin Rutzinger, Kati Heinrich, Axel Borsdorf & Johann Stötter

Content

Scientific output

Avian, M. 2012: First results of repeated Terrestrial Laserscanning monitoring processes at the rock fall area Burgstall / Pasterze Glacier, Hohe Tauern Range, Central Austria. *Geophysical Research Abstracts* 14 (EGU2012-8905).

Avian, M. 2012: Performance and limits of different long-range TLS-sensors for monitoring high mountain geomorphic processes at different spatial scales. *Geophysical Research Abstracts* 14 (EGU2012-10081).

Avian, M. & A. Kellerer-Pirklbauer 2012: Modelling of potential permafrost distribution during the Younger Dryas, the Little Ice Age and at present in the Reisseck Mountains, Hohe Tauern Range, Austria. *Austrian Journal of Earth Sciences* 105, 1: 140–153.

Bollmann, E., J. Abermann, C. Klug, R. Sailer & J. Stötter 2012: Quantifying Rock glacier Creep using Airborne Laserscanning. A case study from two Rock glaciers in the Austrian Alps. In: Hinkel, K.M. (ed.): *Proceedings of the Tenth International Conference on Permafrost,* Salekhard, Russia, 25–29 June 2012. Vol. 1: International Contributions: 49–54.

Bollmann, E., R. Sailer, C. Briese, J. Stötter & P. Fritzmann 2010: Potential of airborne laser scanning for geomorphologic feature and process detection and quantification in high alpine mountains. *Zeitschrift für Geomorphologie* 55, Suppl. 2: 83–104.

Eisank, J. 2012: *Untersuchungen zur Verbreitung von Permafrost im Schrankar, Stubaier Alpen.* Master thesis. Department of Geography and Regional Science, University of Graz.

Kaufmann, V. 2012: Detection and quantification of rock glacier creep using high-resolution orthoimages of virtual globes. *International Archives of the Photogrammetry, Remote Sensing and Spatial Information Sciences* XXXIX-B5: 517–522.

Kellerer-Pirklbauer, A. 2011: Thermal regime of ground surfaces in different alpine areas of Central and Eastern Austria between 2006 and 2010. *Geophysical Research Abstracts* 13 (EGU2011-12999).

Kellerer-Pirklbauer, A. 2011: Freeze-thaw cycles and frost shattering potential at alpine rockwalls in the Hohe and Niedere Tauern Ranges, Austria) between 2006 and 2010. *Geophysical Research Abstracts* 13 (EGU2011-13091).

Kellerer-Pirklbauer, A. 2013: Ground surface temperature and permafrost evolution in the Hohe Tauern National Park, Austria, between 2006 and 2012. Signals of a warming climate? *5th Symposium for Research in Protected Areas – Conference Volume,* Mittersill, Austria: 363–372.

Kellerer-Pirklbauer, A. & M. Avian 2012: Permafrost und Bodentemperaturveränderungen zwischen 2006 und 2011 in der Reißeckgruppe, Hohe Tauern, Österreich. *Carinthia* II 202: 505–522.

Kellerer-Pirklbauer, A., M. Avian, V. Kaufmann, E. Nieser & B. Kühnast 2012: Climate-induced spatiotemporal changes of rock glacier kinematics and temperature regime of permafrost in the Hohe Tauern Range, Austria: One work package within the permAfrost project. *Geophysical Research Abstracts* 14 (EGU2012-9492).

Kellerer-Pirklbauer, A. & V. Kaufmann 2012: On the relationship between surface flow velocity and climatic conditions at three rock glaciers in central Austria. *Geophysical Research Abstracts* 14 (EGU2012-9573).

Kellerer-Pirklbauer, A. & V. Kaufmann 2012: About the relationship between rock glacier velocity and climate parameters in central Austria. *Austrian Journal of Earth Sciences* 105, 2: 94–112.

Kellerer-Pirklbauer, A., G.K. Lieb & H. Kleinferchner 2010: A new rock glacier inventory at the eastern margin of the European Alps. *Geophysical Research Abstracts* 12 (EGU2010-13110).

Klug, C. 2011: *Blockgletscherbewegungen in den Stubaier (Reichenkar) und Ötztaler (Äußeren Hochebenkar) Alpen – Methodenkombination aus digitaler Photogrammetrie und Airborne Laserscanning.* Master thesis. Institute of Geography, University of Innsbruck.

Klug, C., E. Bollmann, K. Krainer, L. Rieg, R. Sailer, M. Sproß & J. Stötter 2011: Combination of photogrammetry and airborne laser scanning to derive horizontal flow velocities and volume changes of rock glaciers. *Geophysical Research Abstracts* 13 (EGU2011-12063-1).

Klug, C., E. Bollmann, R. Sailer, J. Stötter, K. Krainer & A. Kääb 2012: Monitoring of permafrost creep on two rock glaciers in the Austrian Eastern Alps: Combination of aerophotogrammetry and airborne laserscanning validated by dGPS measurements. In: Hinkel, K.M. (ed.): *Proceedings of the Tenth International Conference on Permafrost,* Salekhard, Russia, 25–29 June 2012. Vol. 1: International Contributions: 215–220.

Krainer, K. 2010: Geologie und Geomorphologie von Obergurgl und Umgebung. In: Koch, E.-M. & B. Erschbamer (eds.): *Glaziale und Periglaziale Lebensräume im Raum Obergurgl.* Innsbruck: 31–52.

Krainer, K. & M. Ribis 2010: Blockgletscherinventar Ötztaler – Stubaier Alpen. In: *Permafrost-Workshop Obergurgl.* Obergurgl, Austria, 14–15 October 2010. GeoAlp 7: 100.

Krainer, K., A. Kellerer-Pirklbauer, V. Kaufmann, G.K. Lieb, L. Schrott & H. Hausmann 2012: Permafrost Research in Austria: History and recent advances. *Austrian Journal of Earth Sciences* 105, 2: 2–11.

Rieg, L., R. Sailer, J. Stötter & D. Burger 2012: Vegetation Cover on Alpine Rock Glaciers in Relation to Surface Velocity and Substrate. In: Hinkel, K.M. (ed.): *Proceedings of the Tenth International Conference on Permafrost,* Salekhard, Russia, 25–29 June 2012. Vol. 1: International Contributions: 329–334.

Sailer, R., E. Bollmann, S. Hoinkes, L. Rieg, J. Stötter & M. Sproß 2012: Quantification of geomorphodynamic processes in glaciated and recently deglaciated terrain based on airborne laser scanning data. *Geographiska Annaler,* Series A – Physical Geography 94: 17–32.

Schretter, I. 2012: *Rock Glaciers near the Franz-Senn-Hütte, Oberberg Valley, Stubai Alps, Austria.* Master thesis. Institute of Geology and Paleontology, University of Innsbruck.

Stocker, K. 2012: *Geologie und Blockgletscher der Madererspitze, Vorarlberg.* Master thesis. Institute of Geology and Paleontology, University of Innsbruck.

Watzdorf, S. 2012: *Rock Glaciers in the Samnaun Mountain Group, Western Tyrol, Austria.* Master thesis. Institute of Geology and Paleontology, University of Innsbruck.

Permafrost-Glacier Interaction – Process Understanding of Permafrost Reformation and Degradation

Jan-Christoph Otto [1] **& Markus Keuschnig** [1,2]

[1] *Department of Geography and Geology, University of Salzburg, Salzburg, Austria*
[2] *alpS – Centre for Climate Change Adaptation, Innsbruck, Austria*

1   Introduction

The dramatic changes of mountain glaciers and significant rock fall events during exceptional warm summers in the last decades have strongly raised awareness and interest in changing geomorphologic conditions of high mountain areas. Alpine areas are considered to be particularly sensitive to climate change and observations as well as projections report a rise of temperatures significantly above lowland areas (Bogataj 2007). Temperature increase in high mountain areas affects glacier and permafrost distribution and causes reactions on geomorphological as well as hydrological conditions. Most permafrost areas in high mountains are located in close vicinity to glaciers due to similar environmental requirements. The strong loss of length and volume of Alpine glaciers represent the most visible manifestation of cryosphere change in high mountains. While glacier changes become apparent in relatively short reaction times, mountain permafrost reacts also sensitive to warming but somewhat delayed and almost invisible. The major objective of this study is to investigate the condition and evolution of the ground thermal regimes in glacial and periglacial environments after glacier melt.

The scientific communities of glacier and permafrost research have operated separately in the past, even though, interactions between glaciers and permafrost are recognised (Haeberli 2005). Many equilibrium lines of Alpine glaciers in continental climates are located within zones of permafrost occurrence (Haeberli & Gruber 2008). Thus, the thermal regimes of surface ice and frozen ground can be interconnected influencing each other. Glaciers may exhibit cold or polythermal conditions at the base mainly as a function of energy and mass balance at the surface or influence of negative temperatures from below due to the existence of permafrost (Suter et al. 2001). In the Swiss Alps cold based glacier occurrence is assumed to be restricted to altitudes above 3,800 m (Haeberli 1976; Suter 2001). The occurrence of hanging glaciers and ice patches on steep bedrock slopes of the highest peaks is associated with cold based conditions and the occurrence of permafrost (Haeberli 2005). Disappearing hanging glaciers and ice covered steep slopes during the last century may be the result of warming subsurface conditions within the steep rock walls. However, little is known on this relationship due to scarce data on bedrock permafrost or ice wall thermal conditions (Ravanel & Deline 2011).

Glacier retreat has released significant areas since the last glacier maximum during the Little Ice Age (mid-19th century). With the continuing melt of Alpine glaciers significant space is released at altitudes potentially susceptible for permafrost existence. This space is either located in front of the glacier (forefield) due to length reduction of the glacier or surrounding the glacier due to reduction of glacier thickness. In these areas various polygenetic ground ice occurrences have been observed. The origin of the ice has been assigned to three processes: (Type I) refreezing of former unfrozen glacier beds (i. e. formation of permafrost), (Type II) preservation of previous subglacial permafrost and (Type III) burial of dead ice (Kaab & Kneisel 2006; Kneisel 2003; Kneisel & Kaab 2007; Lugon et al. 2004). However, little is known on the time required for formation of permafrost in Alpine environments (Lunardini 1995) or the preservation of permafrost below glacier coverage. The interpretation of differing observations concerning permafrost thawing and degradation and potential natural hazards (e. g. rock falls, debris flows) remains a major challenge (Haeberli et al. 2010). Efficient risk analysis and risk adaptation strategies depend largely on process understanding of permafrost-related evolution and related hazards. Permafrost degradation is one potential effect of warming trends in the Alps (APCC 2014) leading to destabilisation of bedrock slopes and increased potential of debris slow generation (Sattler et al. 2011). However, in order to assess the future impact of permafrost areas to the formation of natural hazards due to climate change, knowledge of the glacier-permafrost interaction is required. This includes understanding of the different reaction times of glaciers and permafrost zones to temperature increase. If glacier melt happens faster than subsurface warming we could experience an incre-

Figure 1: View of the Schmiedingerkees glacier below Kitzsteinhorn peak (center left)

ase in permafrost area in high Alpine terrain previously covered by glacier ice. This would also increase the hazard potential in these areas and needs to be considered for planning of adaptation strategies (Keuschnig et al. 2011).

This study aims at understanding the permafrost-glacier relationship in the Kitzsteinhorn area, Kaprun, Austria (Fig. 1). By analysing both the recent history of glacier ice change and the current occurrence of permafrost and its thermal state and conditional parameters (climate, land surface parameters) we aim to understand the existence or permafrost conditions in the direct vicinity of the glacier. The main research questions include:

- What are the ground thermal conditions around the Schmiedingerkees glacier?
- Can we observe and identify permafrost occurrence?
- When did the permafrost locations become exposed from the glacier cover?
- Which factors influence the ground thermal conditions around the Schmiedingerkees glacier?

2   Test site

The study is located at the Schmiedingerkees cirque at the Kitzsteinhorn ski area in the Federal Province of Salzburg, Hohe Tauern Range, Austria. The cirque opens in north-eastern direction from the summit of the Kitzsteinhorn (3,203 m), covering approximately 3 km² and a vertical elevation difference of 1,500 m between the summit and the glacier forefield limits (1,700 m maximum Little Ice Age extent). The Kitzsteinhorn is located just north of the main Alpine divide and has no directly adjacent summits. The Schmiedingerkees glacier has a size of approximately 1.05 km² (2012), covering around 40% of the cirque area. The glacier is a flat cirque type glacier surrounded by steep bedrock slopes of up to 250 m height (Fig. 3).

The Kitzsteinhorn area primarily consists of calcareous mica schists (Höck & Pestal 1994). Stress release and intense physical weathering processes, typical for periglacial environments, resulted in the formation of an abundance of joint sets with large apertures in the rock walls of the peak and adjacent cirque walls. Intense retreat of the Schmiedingerkees glacier in recent decades led to the exposure of oversteepened rock faces, which in turn are frequently affected by minor rock fall events (Hartmeyer et al. 2012). The recently exposed glacier forefield is characterised by large areas of exposed bedrock (Fig. 3). Only lateral and lower parts are debris covered and

Table 1: Climate data of the reference climate stations around the Kitzsteinhorn

Location	Altitude [m]	Time period	MAAT [°C]	Mean snow height [m]	Max snow height [m]	Mean solar radiation [W / m²]
Alpincenter	2,446	01.2005–08.2013	0.78	0.93	2.8	–
Kammerscharte	2,561	11.2008–08.2013	–3.23	1.2	3.6	166.8
Glacier Plateau	2,910	11.2008–08.2013	–2.99	1.5	4.1	–

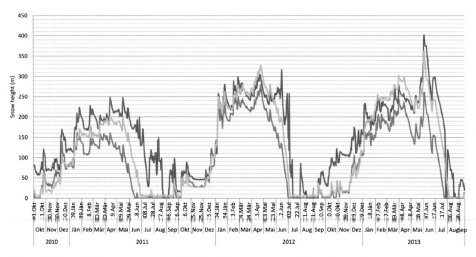

Figure 2: Snow height measurement at the three climate stations in the study area between 2010 and 2013

contain surface indicators for previous glacier extent (lateral, frontal moraines). Apparently, the Schmiedingerkees has a low debris production compared to other glaciers. The largest area of thick debris cover is located in the eastern part of the glacier forefield, east of the Schmiedinger lake. The steep cirque side walls are characterised by intense rock fall and avalanche activity. Especially the eastern ridge, descending from the Kitzsteinhorn peak shows intense erosion leading to debris cover of the eastern part of the glacier. This debris input is most probably responsible for the debris accumulation in the forefield at the eastern side. Prominent lateral moraines that indicate the Little Ice Age (LIA) maximum can be observed south of the Alpincenter (2,446 m) and within the descending valley north of the lake. Three weather stations are located within the study area, permitting continuous observation of external forcing of ground thermal conditions. The weather stations are located at the Alpincenter (2,446 m) of the ski station at the Kammerscharte (2,561 m) in the neighbouring cirque towards the southeast and directly on the Schmiedingerkees glacier (2,940 m). The stations show mean annual air temperature (MAAT) values of 0.8 °C, –3.2 °C and –3.0 °C, respectively. The large variability could be the result of local influences on the measurement, for example the impact of warming from the building at the Alpincenter may result in an increase in MAAT. Maximum snow heights between 2.8 and 4.1 m are recorded between 2005 and 2013 (Table 1, Fig. 2).

The tourism infrastructure existing within the study area (cable car, ski lifts, ski slopes, etc.) provides easy access and convenient transportation of measuring equipment, an essential prerequisite for an extensive long-term monitoring program (Keuschnig et al. 2011). However, the glacier forefield is strongly affected by the intense usage and modification of the terrain for the construction of ski slopes, roads and buildings thus having an impact on debris characteristics and ground thermal condition. Since most of the skiing is performed on the glacier itself, the station ma-

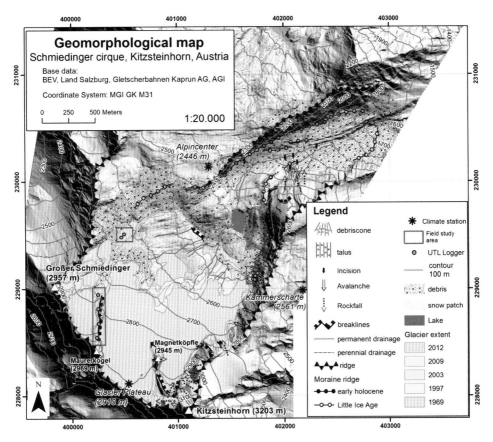

Figure 3: Geomorphological map of the Schmiedingerkees cirque, Kitzsteinhorn, Kaprun, Austria

nagement also involves the glacier conditions by constructions of ski slopes, filling of crevasses and lifts tracks on the ice. In order to minimise direct human impact on the subsurface conditions we chose two locations close to the glacier where little or no construction works or surface modification was performed.

3   Methods

The study combines field data with remote sensing and Geographical Information System (GIS) analysis and is split into research on the changes of the glacier forefield induced by glacier retreat and investigation on the permafrost occurrence and the measurement of surface / subsurface temperatures.

The field work comprises permafrost detection by electrical resistivity tomography (ERT) and measurement of surface / subsurface temperatures using data loggers.

Resistivity measurements were performed using a GeoTomMK8E1000[1] multi-electrode resistivity system with 24 electrodes and 2 to 4 m electrode spacing. ERT was analysed with the Res2DInv software package.

For the collection of ground surface temperature (GST) data we placed ten Universal Temperature Loggers (UTL) (Type UTL1, Geotest.ch, ex-factory accuracy of ±0.1 °C) in the top layer (–5 to –10 cm) of the subsurface. Temperature loggers were placed at three different locations at various altitudes and on different subsurface conditions (fine grain material, coarse grain material, and close to bedrock) and where covered by fine grain sediments to avoid direct exposure to the sun and snow. Additionally, we could use climate data from three climate stations in the Kitzsteinhorn area, recording temperature, precipitation, snow height, solar radiation and wind. All climate and temperature data was stored and analysed using a Microsoft Access database. We derived mean annual ground surface temperature (MAGST), winter equilibrium temperature (WEqT), duration of snow cover (SCD) and estimated time since deglaciation for all GST locations. MAGST is calculated for entire years if available and the entire data set. In case of missing records we added the missing days from neighbouring locations with similar data as previously applied by Apaloo et al. (2012). WEqT is generally considered as stable temperature during the longest continuous duration of thick snow cover (> 50 cm) over a minimum duration of two weeks (Schoeneich 2011). Snow height measurements at the surrounding climate stations indicate a thick snow cover of at least 1 m or more for most of the winter until at least May at wind sheltered locations (Fig. 2). The formation of a WEqT and the interpretation of WEqT conforming to the Bottom Temperature of the Snow cover (BTS) principle should therefore be possible for our logger sites (Schoeneich 2011). WEqT were extracted by visual inspection of the temperature data timelines in the database. Morphometric land surface characteristics have been calculated (slope, aspect, and total insolation) for the logger sites to analyse external location influences. SCD quantification is based on observations made by Schmidt et al. (2009) who identified a standard deviation of less than 0.3 K of GST during 24 h as good indicator of snow coverage. Additionally, we estimated the time since deglaciation based glacier extent visible on the aerial imagery available.

For the GIS and remote sensing analyses different digital elevation models (DEM) and different aerial images have been collected. Geomorphological features have been mapped using airborne laser scanning (ALS) data (Land Salzburg and Gletscherbahnen Kaprun AG) with 1 m resolution and high resolution aerial imagery (2012, Land Salzburg). Data on glacier extends have been generated by mapping on digital orthophotos from 1982, 1997, 2003, 2009 and 2012 (Land Salzburg). Glacier extent from 1969 was extracted from the Austrian Glacier Inventory (Gross 1987). Morphometric landform parameters have been calculated in ArcGIS and SAGA GIS using 1 m ALS DEM and a 5 m x 5 m analysis window to eliminate local derivation.

---

[1]   http://geolog2000.de (17.12.2013)

Figure 4: Historical images (postcards) showing the glacier extend of the Schmiedingerkess below the Kitz-steinhorn. The postcard on the right is marked with 15. August 1906 (Verlag Würthle & Sohn, Salzburg, No. 208), the image on the left is dated to 1933 (Bergwelt Verlag, C. Jurischek, Salzburg; historical images kindly provided by Heinz Slupetzky, Salzburg). The glacier terminus has reached the cirque boundary. The glacier is filling large parts of the cirque and has a connection to the Kammerkees glacier towards the eastern flank of the Kitzsteinhorn peak. Also visible is a pronounced ice cover on the steep slopes of the peak. The Magnetköpf-le, a small peak towards the right of the Kitzsiteinhorn seems to be almost completely ice covered in 1906. In contrast the Maurergrat ridge, visible on the right image in the upper right part of the Schmiedingerkees glacier was only partially covered with ice.

4 Results

4.1 Changes of the Schmiedingerkees glacier area

The Schmiedingerkees has experienced a total loss of around 70% of area covered at the LIA maximum (Table 2). The length change is about 2.4 km since the LIA and 300 m since the onset of length records in 1951 (WGMS 2012). During its LIA maximum the glacier is terminated in a pronounced tongue at an altitude of approximately 1,635 m above Kaprun valley, leaving the cirque area. Based on morphological mapping, the maximum extent of the glacier could be reconstructed. For area calculation it is assumed that the glacier ice filled the cirque to a great portion leaving only higher parts of the surrounding cirque walls free of ice. Since early 20th century the glacier was restricted to the cirque area and changed into a flat cirque glacier with no pronounced glacier tongue (Fig. 4). The glaciers lost an average of 15,000 m² of area per year between LIA and 2012. In the last years (2009–2012), this number has doubled.

The melting of the glacier released an area of 2.4 km² since the LIA at altitudes between 1,635 and 3,200 m. Strongest changes in glacier area are by nature observed in the lower part of the glacier, but significant area is released of ice in the upper parts as well (Fig. 5). Especially the existence of glacier ice on the steep northern rock wall of the Kitzsteinhorn that existed until the 1980s has released significant surface here.

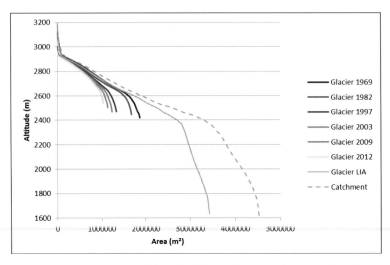

Figure 5: Hypsometric curve of the Schmiedinger glacier area between LIA and 2012. Data based on 1 m DEM (Land Salzburg and Gletscherbahnen Kaprun AG), geomorphological mapping and digital orthophoto analysis

4.2   Permafrost evidences

4.2.1   Ground surface temperature data

The GST loggers have been placed in the eastern and western part of the glacier fore-field as well as on the Maurergrat, a ridge separating the Schmiedingerkees from the Maurerkees in the west (Fig. 3). In the glacier forefield east, loggers are located on a steep talus deposit of fine to coarse grain size (unfortunately these logger only recorded data from one hydrological year due to technical failure). The loggers are placed at altitudes between 2,534 m and 2,546 m within a distance of 30 m. The location is assumed to be free of glacier ice a maximum of 40 years (Table 3). But it is like-

Table 2: Glacier area changes of the Schmiedingerkees based on geomorphological mapping and orthophoto interpretation

Year	Area [km²]	Change to previous date [km²]	%	Change to LIA maximum [km²]	%
LIA (assumed 1850)	3.4	0	0	0	0
1969	1.88	−1.5	44.7	−1.5	−44.7
1982	1.69	−0.2	9.9	−1.7	−50.2
1997	1.34	−0.4	20.7	−2.1	−60.5
2003	1.24	−0.1	7.6	−2.2	−63.5
2009	1.15	−0.1	7.6	−2.3	−66.3
2012	1.05	−0.1	8.6	−2.4	−69.2

ly that this slope previously contained remains of the debris covered glacier tongue until a few years ago. Impressive stripes of the debris, visible on the aerial images, correspond to the previous movement of the glacier ice. In the western part of the glacier forefield loggers are placed on little inclined terrain in small pockets of fine sediments between polished bedrock outcrops at altitudes of 2,631 m and 2,623 m. This terrain is assumed to be free of ice since 15 to 30 years based on the aerial images. The GST loggers located on the Maurergrat ridge also placed in small pockets

Table 3: Land surface parameters of the GST data loggers

Location	Recording period	Altitude [m]	Slope [°]	Aspect [°]	Surface cover	Rugged-ness Index	Total Insolation per year [kWh / m^2]	Estimated time since deglacia-tion
Glacier forefield East								
UTL-2087	09.11–07.12	2,538	39.8	303.5	Talus slope, fine grain sediment, close to bedrock	0.59	1,278.1	Max.40
UTL-707	09.11–02.13	2,534	37.1	289.5	Talus slope, fine grain sediment	0.53	1,442.3	Max.40
UTL-759	09.11–12.12	2,546	38.4	297.0	Talus slope, fine grain sediment, close to bedrock	0.56	1,354.9	Max.40
UTL-702	09.11–07.12	2,537	37.2	306.7	Talus slope, fine grain sediment	0.54	1,293.0	Max. 40
Glacier forefield West								
UTL-2104	09.09–10.12	2,631	18.6	124.1	glacier forefield, medium grain sediment, close to bedrock	0.24	2,227.1	15–30
UTL-2092	09.09–10.12	2,623	17.1	15.5	glacier forefield, medium grain sediment, close to bedrock	0.23	1,531.6	15–30
Maurergrat								
UTL- 2067	09.09– 9.12	2,915	14.5	336.0	Ridge, fine grain sediment, close to bedrock	0.18	1,794.4	Max.44
UTL- 2091	09.09–10.12	2,878	2.1	261.8	Ridge, fine grain sediment, close to bedrock	0.14	2,139.9	Max. 44
UTL-2095	09.09–05.12	2,847	19.1	302.3	Ridge, fine grain sediment, close to bedrock	0.25	1,811.5	Max. 44
UTL-2074	09.09–09.12	2,775	49.5	316.5	Ridge, fine grain sediment, close to bedrock	0.84	958.1	Max. 44

Table 4: Ground thermal data of the logger sites

Location	Altitude [m]	MAGST [°C]	Time period for MAGST	WEqT [°C] [year]	Mean duration of snow cover [days]
Glacier forefield east					
UTL-2087	2,538	1.02	09.2011–09.2012	−1.3 [2011]	224
UTL-707	2,534	0.00	09.2011–09.2012	−2.3 [2011]	225
UTL-759	2,546	1.04	09.2011–09.2012	−2.3 [2011]	245
UTL-702	2,537	1.40	09.2011–09.2012	−1.2 [2011]	255
Glacier forefield west					
UTL-2104	2,631	2.12	09.2009–09.2012	−1.2 [2012]	245
UTL-2092	2,623	0.72	09.2009–09.2012	−1.9 [2012]	279
Maurergrat ridge					
UTL- 2067	2,915	−0.52	09.2009–09.2011	−3.8 [2012]	302
UTL- 2091	2,878	−1.17	09.2009–09.2011	−4.1 [2012]	315
UTL-2095	2,847	−1.71	09.2009–09.2011	−3.4 [2010]	124.5
UTL-2074	2,775	−1.58	09.2009–09.2011	−4.5 [2010]	222

of fine sediment in close proximity to the bedrock outcrop. These loggers are placed along an altitudinal transect just of the western side of the ridge at altitudes between 2,775 m and 2,915 m. The loggers are approximately 130 m apart from each other. The western part of the ridge has been covered by glacier ice observable on the aerial photos of 1969. Apparently the ridge was never completely ice covered (see Fig. 4 right). It is assumed that the bedrock of the east facing rock wall was free of ice during LIA maximum extent.

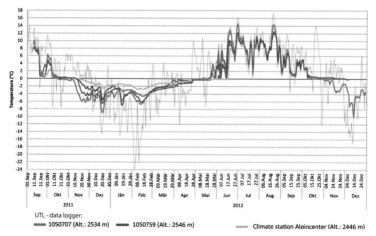

Figure 6: GST recorded between Sept. 2011 and Dec. 2013 at Glacier Forefield East and air temperature measured at the Alpincenter climate station

The MAGST for all logger locations is presented in Table 4. The loggers placed along the Maurergrat ridge have MAGST values below zero indicating potential permafrost conditions, at all other locations MAGST is at or above zero degrees. A closer look at the annual variation in GST is presented in Figures 6 to 8. At the Glacier Forefield East (GFE) all loggers show a typical early winter temperature variation following roughly daily temperature changes (Fig. 6). Snow cover starts to develop in early October and lasts until end of May observable by the zero curtain

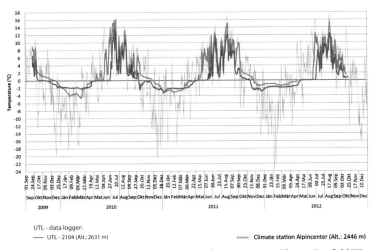

Figure 7: GST recorded between Sept. 2009 and Oct. 2012 at Glacier Forefield West and temperature data measured at the Alpincenter climate station

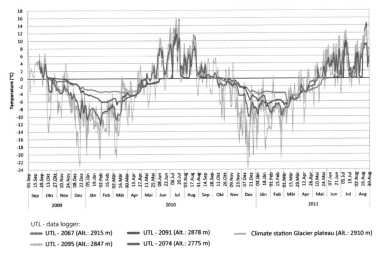

Figure 8: GST recorded between Sept. 2009 and Aug. 2011 at Maurergrat ridge and temperature data measured at the glacier plateau climate station

effect in the data, corresponding to snow height measurements at the neighbouring climate stations. However, only two loggers have smooth winter curves (UTL 2087, UTL 1050702) indicating a better isolation effect of the snow cover here that allows for a development of a WEqT. We extracted a WEqT at these locations between −1.2 °C and −2.3 °C indicating that this is a boundary location where permafrost is possible.

At the Glacier Forefield West (GFW) we have three years of winter recording (2009 to 2011) and smoother winter curves compared to the situation at glacier forefield east (Fig. 7). Both loggers reveal very stable temperature conditions indicating a thick permanent snow cover and little impact from air temperature. Onset of snow cover lies between the 8th and 17th of October between 2009 and 2011. It is observable that the general trends between the lines change from year to year with generally colder temperatures at UTL 2104 (blue curve) in winter 2010 and parts of winter 2011 and a higher temperatures in 2012. This could be related to the effect of snow cover at the two sites. At site UTL 2091 (red curve) the late winter zero curtain is much longer compared to the neighbouring site and lasts till mid-June. This indicates a thicker snow cover that may also be responsible for quite stable GST values between January and April. This location seems to be better sheltered protecting the snow cover from wind and sun more than the other location nearby. The long period of zero curtain effect could also be responsible for the significant lower MAGST compared to site UTL 2104. Though MAGST is positive, WEqT of −1.9 °C indicates that this location has a weak potential to provide permafrost conditions.

At the Maurergrat ridge GST is recorded since 2009 and the last data was gathered in 2011. Here, two locations (UTL 2067 and UTL 2091) show smooth winter curves compared to the other two locations that show strong variations during winter (Fig. 8). The latter locations (UTL 2095 and UTL 2074) seem to have less thick snow cover recognisable in missing of a pronounced zero curtain effect at the end of the winter. Since the loggers are located close to the ridge it is very likely that wind erosion of snow play a major role here. All loggers show MAGST temperatures between −0.5 and −1.7 °C giving a clear indication for permafrost conditions. Since measurement conditions at UTL 2095 and UTL 2074 seem to be strongly affected by wind activity leading to a removal of the isolating snow cover, determination of WEqT is difficult at these sites. We therefore only discuss WEqT at the sites UTL 2067 and UTL 2091. At these two locations WEqT of −3.8 and −4.0 °C, respectively, are clear indicators of permafrost condition.

4.2.2   Resistivity data
Resistivity measurements have been performed at various locations in the glacier forefield and on the Maurergrat ridge (Fig. 3). Figure 9 depicts the resistivity conditions at GFE within a few meters to the actual glacier terminus. The profile runs from the debris covered glacier tongue (left) towards north-east into the proglacial debris (right). The resistivity values are between 1,000 and >1,000,000 Ωm. The resistivity distribution clearly marks the transition between the glacier ice with values above 100,000 Ωm and the non-frozen zone in the proglacial area with values

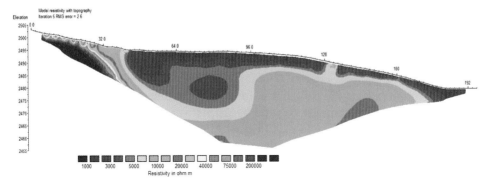

Figure 9: Resistivity measurement at GFE

< 20,000 Ωm. Two other measurements at this part of the glacier forefield produced a similar image. Based on resistivity measurements it is unlikely that permafrost conditions are present in the eastern part of the deglaciated area at an altitude of 2,490 to 2,530 m on a northeast exposed slope.

In contrast, clear permafrost evidence by high resistivity values can be observed on the ridge of the Maurerkogel at altitudes between 2,875 and 2,950 m (Fig. 10). The measurement reveals a clear horizontal layering of resistivity values with a distinct rise above 20,000 Ωm in about 5 to 8 m depth. Resistivity data here depicts a typical Alpine late summer permafrost situation (date of measurement Sept. 2009) with an unfrozen active layer (resistivity < 20,000 Ωm) and permafrost conditions indicated by resistivity values of >10,000 Ωm. This ERT profile runs parallel to the location of the GST loggers at the ridge and backs up the GST observation.

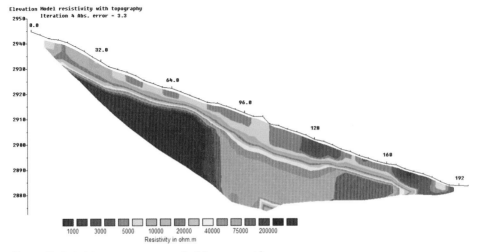

Figure 10: Resistivity measurement on the Maurergrat ridge

4.2.3 Permafrost modelling

A statistical model of permafrost distribution (Permakart 3.0) has been applied for the Kitzsteinhorn area based on a preceding study conducted by the authors at the University of Salzburg (Permalp.at project). The model is based on empirical permafrost data from the Hohe Tauern range (Schrott et al. 2012) and presents an index of probability of permafrost occurrence. An area of 1.2 km² within the cirque of the Schmiedingerkees is potentially covered by permafrost. The model shows that large parts of the current glacier forefield lies within the potential zone for permafrost (Fig. 11). The lowest potential permafrost zones are located on steep, northern exposed slopes or isolated patches. Below 2,600 m front of the Schmiedingerkees glacier and below 2,700 m in front of the Maurerkees glacier, the probability of permafrost occurrence drops below 20%. Below 2,500 m only very isolated patches provide permafrost conditions. Both test sites in the forefield lie outside the modelled permafrost area, but are very close to lower limit modelled.

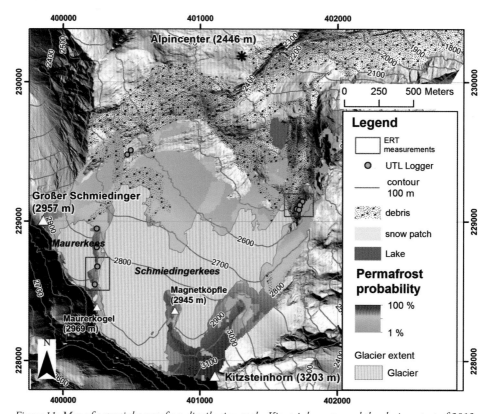

Figure 11: Map of potential permafrost distribution at the Kitzsteinhorn around the glacier extent of 2012

4.3   Discussion

Based on the permafrost model and the mapped glacier area we can identify which parts of the areas exposed by glacier melt are potentially under permafrost conditions. Assuming the same distribution of permafrost 40 years ago the potential area of permafrost condition increased from 0.5 to 1.1 km² between 1969 and 2012 by the melting of glacier ice. Looking at the altitudinal distribution of permafrost area we notice a very homogenous increase of permafrost area across the entire spread of the cirque (Fig. 12). This is related to the overall decreasing glacier thickness and the release of rock walls alongside the margins of the glacier. Additionally, the former ice cover of the Kitzsteinhorn north face produces a slightly stronger increase at the highest altitudes above 2,950 m.

While indication on permafrost presence is strong at the Maurergrat ridge we found less likely evidence for permafrost in the glacier forefield. At the Maurergrat ridge both ERT and GST data show permafrost occurrence, which is also modelled by the permafrost distribution model (Fig. 11). The ERT profile shows a pronounced permafrost body with an unfrozen top layer between 5 and 8 m depth. WEqT < –3 indicates that this top layer is refreezing during winter representing an active permafrost occurrence. ERT values show a permafrost thickness of at least 20 m below the active layer (Fig. 10). We assume that this ridge was most probably ice covered on the top during the LIA, but the eastern rock wall was still exposed and mostly ice free. We interpret this permafrost occurrence to be a preserved, pre-existing ground ice. Under these assumptions the glacier ice of the Schmiedingerkees must have been cold based when it covered parts of the Maurergrat ridge. The ground thermal conditions at this site are most likely influenced by three dimensional effects from the adjacent rock wall where negative temperature impact penetrated into the bedrock

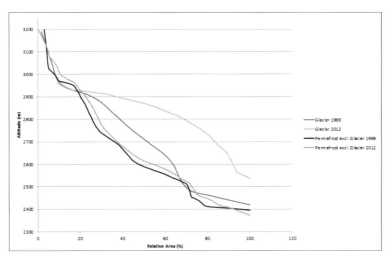

Figure 12: Hypsometry distribution of glacier ice and potential permafrost area between 1969 and 2012

ridge. Currently, the strong disturbances of the snow cover due to the exposed location and strong wind contributes to negative energy balances during most of the winter preserving the permafrost condition within the ridge.

At GFE we did not find permafrost indication in the ERT measurements. However, we assumed that until a few decades ago this slope was covered by preserved debris covered ice, similar to the current conditions only 30 m upwards. This ground ice seems to be completely vanished today. GST values however indicate a weak potential for permafrost conditions at the upper part of the slope in transition to the bedrock. This is revealed also by the permafrost model. Very low annual total radiation values below 1,400 kWh/yr. may be favouring permafrost development here. It is possible that the time period of around 40 years for permafrost formation has been too short at this location or that MAAT at this altitude is too high for formation of permafrost.

Looking at the data from GFW we can conclude that permafrost conditions are possible based on ground thermal conditions, but not verified by additional measurements. This location is less steep and receives a stronger insolation input compared to the eastern glacier forefield (Table 3). Thus, a formation of new permafrost conditions is less likely here compared to the eastern glacier forefield.

5   Conclusions

A comparison of glacier area change and permafrost distribution modeling shows that significant space has been exposed with permafrost conditions between 2,400 and 3,200 m. Additionally, we could observe a significant negative ground thermal regime indicating permafrost conditions at a ridge location between 2,770 and 2,910 m that has been partially ice covered in the past. Due to the thickness of the permafrost layer we classify this permafrost occurrence as preserved ground ice that has been in place for a long time. Current local environmental conditions contribute to the preservation of this ground ice today. In the glacier forefield WEqT data indicate a possibility for permafrost, but additional data form ERT does not reveal permafrost existence. We thus cannot identify new formation of permafrost at the glacier forefield, which is either due to too little time for formation or due to too strong positive energy input at these altitudes. A continuing of GST measurements at this boundary location is required before information on permafrost formation and the required time period is available. We therefore propose an ongoing monitoring to gain further insight into this sensitive land surface condition.

6   Outlook

In order to better understand the preservation and possible formation of permafrost conditions in glacier forefield longer time series of ground thermal data are required.

This enables to evaluate the consequences of ground thermal regime changes after surface exposure by ice melt and helps to understand the time frame at which new permafrost is build up. Based on the observations at the Schmiedingerkees glacier area we have to conclude that a large part of these potential sensitive zones are located in very steep terrain with limited accessibility. Ongoing monitoring should therefore benefit from existing logistical support from cable cars and existing infrastructure for data collection despite more human impact and disturbance at these sites.

7   Acknowledgements

This study is funded by the Austrian Academy of Science (ÖAW). This support is gratefully acknowledged. Furthermore, this investigation benefits from the close cooperation with the alpS project MOREXPERT, the logistic and financial support by the Gletscherbahnen Kaprun AG and the highly valuable help by colleagues and students in the field. We thank all supporters for their help.

8   References

Apaloo, J., A. Brenning & X. Bodin 2012: Interactions between Seasonal Snow Cover, Ground Surface Temperature and Topography (Andes of Santiago, Chile, 33.5° S). *Permafrost and Periglacial Processes* 23, 4: 277–291.

APCC (in prep.): *Österreichischer Sachstandsbericht Klimaänderungen 2014 (AAR14)*. Climate Change Centre Austria.

Bogataj, L.K. 2007: How will the Alps respond to climate change? Szenarios for the future of Alpine water. In: Psenner, R. & R. Lackner (eds.): *The Water Balance of the Alps – What do we need to protect the water resources of the Alps? Proceedings of the Conference held at Innsbruck University, 28–29 September 2006*. Innsbruck: 43–51.

Haeberli, W. 1976: Eistemperaturen in den Alpen. *Zeitschrift für Gletscherkunde und Glazialgeologie* 11, 2: 203–220.

Haeberli, W. 2005. Investigating glacier – permafrost relationships in high-mountain areas: historical background, selected examples and research needs. In: Harris, C. & J.B. Murton (eds.): *Cryospheric Systems – Glaciers and Permafrost*. Geological Society Special Publication 242. London: 29–37.

Haeberli, W. & S. Gruber 2008: Research challenges for permafrost in steep and cold terrain: an alpine perspective. In: Kane, D.L. & K.M. Hinkel (eds.): *Proceedings of the Ninth International Conference on Permafrost*, Fairbanks, Alaska, 29 June–3 July 2008. Vol. 1: 597–605.

Haeberli, W., J. Noetzli, L. Arenson, R. Delaloye, I. Gärtner-Roer, S. Gruber, K. Isaksen, C. Kneisel, M. Krautblatter & M. Phillips 2010: Mountain permafrost: development and challenges of a young research field. *Journal of Glaciology* 56, 200: 1043–1058.

Hartmeyer, I., M. Keuschnig & L. Schrott 2012: Long-term monitoring of permafrost-affected rock faces – A scale-oriented approach for the investigation of ground thermal conditions in alpine terrain, Kitzsteinhorn, Austria. *Austrian Journal of Earth Science* 105, 2: 128–139.

Höck, V. & G. Pestal 1994: *Geological map of Austria 1 : 50,000 Sheet 153, Großglockner*. Geologische Bundesanstalt Vienna.

Kaab, A. & C. Kneisel 2006: Permafrost creep within a recently deglaciated glacier forefield: Muragl, Swiss Alps. *Permafrost and Periglacial Processes* 17, 1: 79–85.

Keuschnig, M., I. Hartmeyer, J.-C. Otto & L. Schrott 2011: A new permafrost and mass movement monitoring test site in the Eastern Alps – Concept and first results of the MOREXPERT project. In: Borsdorf, A., J. Stötter & E. Veulliet (eds.): *Managing Alpine Future II – Inspire and drive sustainable mountain regions, Proceedings of the Innsbruck Conference, November 21–23, 2011.* IGF-Forschungsberichte 4. Innsbruck: 163–173.

Kneisel, C. 2003: Permafrost in recently deglaciated glacier forefields measurements and observations in the eastern Swiss Alps and northern Sweden. *Zeitschrift Fur Geomorphologie* 47, 3: 289–305.

Kneisel, C. & A. Kääb 2007: Mountain permafrost dynamics within a recently exposed glacier forefield inferred by a combined geomorphological, geophysical and photogrammetrical approach. *Earth Surface Processes and Landforms* 32, 12: 1797–1810.

Lugon, R., R. Delaloye, E. Serrano, E. Reynard, C. Lambiel & J.J. Gonzalez-Trueba 2004: Permafrost and Little Ice Age glacier relationships, Posets Massif, Central Pyrenees, Spain. *Permafrost and Periglacial Processes* 15, 3: 207–220.

Lunardini, V.J. 1995: Permafrost Formation Time. In: US Army Corps of Engineers – Cold Regions Research and Engineering Laboratory (ed.): *CRREL Report* 95-8.

Ravanel, L. & P. Deline 2011: Climate influence on rockfalls in high-Alpine steep rockwalls: The north side of the Aiguilles de Chamonix (Mont Blanc massif) since the end of the 'Little Ice Age'. *The Holocene* 21, 2: 357–365.

Sattler, K., M. Keiler, A. Zischg & L. Schrott 2011: On the Connection between Debris Flow Activity and Permafrost Degradation: A Case Study from the Schnalstal, South Tyrolean Alps, Italy. *Permafrost and Periglacial Processes* 22, 3: 254–265.

Schmidt, S., B. Weber & M. Winiger 2009. Analyses of seasonal snow disappearance in an alpine valley from micro- to meso-scale (Loetschental, Switzerland). *Hydrological Processes* 23, 7: 1041–1051.

Schoeneich, P. 2011: GST – Ground surface Temperature, Permanet Guidelines for Monitoring. http://www.permanet-alpinespace.eu/archive/pdf/GST.pdf (accessed: 3.3.2014).

Schrott, L., J.-C. Otto & F. Keller 2012: Modelling alpine permafrost distribution in the Hohe Tauern region, Austria. *Austrian Journal of Earth Science* 105, 2: 169–183.

Suter, S. 2001: *Cold Firn and Ice in the Monte Rosa and Mont Blanc Areas: Spatial Occurrence, Surface Energy Balance and Climatic Evidence.* PhD thesis. Swiss Federal Institute of Technology Zurich ETH Zürich.

Suter, S., M. Laternser, W. Haeberli, R. Frauenfelder & M. Hoelzle 2001: Cold firn and ice of high-altitude glaciers in the Alps: measurements and distribution modelling. *Journal of Glaciology* 47, 156: 85–96.

WGMS 2012: Fluctuations of Glaciers 2005–2010 (Vol. X). In: Zemp, M., H. Frey, I. Gärtner-Roer, S.U. Nussbaumer, M. Hoelzle, F. Paul & W. Haeberli (eds.): *ICSU (WDS)/IUGG (IACS)/UNEP/UNESCO/WMO, World Glacier Monitoring Service, Zürich, Switzerland.* Based on database version doi:10.5904/wgms-fog-2012-11.

Climatic-induced spatio-temporal change of kinematics and ground temperature of rock glaciers and permafrost in the Hohe Tauern Range, Austria

Andreas Kellerer-Pirklbauer[1,2], **Michael Avian**[1], **Viktor Kaufmann**[1], **Erich Niesner**[3†] **& Birgit Kühnast**[4]

[1] *Institute of Remote Sensing and Photogrammetry, Graz University of Technology, Austria*
[2] *Department of Geography and Regional Science, University of Graz*
[3] *Department of Applied Geological Sciences and Geophysics, University of Leoben, Austria*
[4] *KNGeolektrik e.U., Leoben, Austria*

1 Introduction

High altitude and high latitude regions are generally recognized as being particularly sensitive to the effects of the ongoing climate change (e. g. French 1996, Haeberli et al 1993). A large part of permafrost, permafrost-related active rock glaciers and glaciers in the European Alps are for instance at or close to melting conditions and therefore very sensitive to degradation or to disappearance caused by atmospheric warming. Knowledge regarding permafrost distribution and its climatologically driven dynamics in the entire European Alps is still far from being complete although promising modelling approaches for this scale exist (Boeckli et al. 2012).

Active rock glaciers are creep phenomena of continuous and discontinuous permafrost in high-relief environments moving slowly downvalley or downslope (Barsch 1996; Haeberli et al. 2006; Berthling 2011). Rock glaciers are often characterised by distinct flow structures with ridges and furrows at the surface with some similarity to the surface of pahoehoe lava flows. At steeper parts or at the front of a rock glacier, the rock glacier body might start to disintegrate (e. g. Avian et al. 2009) or even completely tear apart and collapse (Krysiecki et al. 2008). Over time and during climate warming, an active rock glacier might turn first to inactive (widespread permafrost, no movement), second to pseudo-relict (sporadic to isolated permafrost, no movement) and finally to relict (no permafrost, no movement) (Barsch 1996; Kellerer-Pirklbauer 2008).

The projects ALPCHANGE (2006–2011; funded by the Austrian Science Fund FWF) and PermaNET (2008–2011; co-funded by the European Union within the Alpine Space framework) formed major basis in terms of data and expertise for the permAfrost project. Within ALPCHANGE a comprehensive monitoring network was established in the Hohe Tauern Range, Austria. These devices operate successfully since summer 2006 and delivered promising results since then. Furthermore and with regards to content, important long-term monitoring activities such as (i) geodetic surveys of particular rock glaciers (started in 1995), (ii) terrestrial laser scanning (TLS) of a rock glacier (started in 2000) and rock walls in permafrost conditions (started in 2009) as well as (iii) monitoring of the thermal regime of the ground (started 2006) were continued within the permAfrost project.

Research within permAfrost aimed to continue and improve previously carried out research in the field of kinematics, volumetric and thermal monitoring of rock glacier and permafrost and to understand the inner structure of rock glaciers applying a combination of geophysical methods. Thereby, research focused particularly on the rock glaciers Weissenkar (WEI), Hinteres Langtalkar (HLC) and Dösen (DOE) all located in the Hohe Tauern Range, Austria. These rock glaciers are of special interest for rock glacier research because they are one of the best studied rock glaciers in Austria but also to some extent of the European Alps. Furthermore, data and expertise gathered during the project period 2010 to 2013 from other study areas in the Hohe Tauern Range were also used in this study for result optimization. Therefore the aim of this project was to provide a deeper insight into kinematics, morphodynamics and thermal state of the three rock glaciers and their close vicinity by using a multidisciplinary approach applying geodesy, aerial photogrammetry, TLS, geophysical techniques and automatic monitoring.

2   Study areas

The three studied rock glaciers WEI, HLC and DOE are all located in the Tauern Range in Austria. The Tauern Range is an extensive mountain range in the central part of the Eastern Alps covering 9,500 km² in Austria and Italy and is commonly separated into the Hohe Tauern Range and the smaller Niedere Tauern Range. The former covers ca. 6,000 km² and reaches with Mt. Großglockner 3,798 m a. s. l. the highest summit of Austria. As mentioned above, this research project focuses on three different rock glaciers in the Hohe Tauern Range, two of them are located in the sub-unit Schober Mountains and one in the sub-unit Ankogel Mountains. Figure 1 gives an overview about the studied rock glaciers. Figure 2 shows detailed maps of the three rock glaciers with spatial information about instrumentation and measurements.

The three rock glaciers can be briefly characterised as further below. For detailed description and overview about previous research at these rock glaciers see Kellerer-Pirklbauer and Kaufmann (2012).

Weissenkar Rock Glacier (WEI): N 46° 57', E 12° 45'; elevation range from 2,620 to 2,870 m a. s. l. with a length of 650 m and a width of 300 m. WEI is a slowly moving tongue-shaped rock glacier consisting of an upper lobe overriding a lower lobe and characterized by well developed furrows and ridges at the entire lower half of the rock glacier. WEI moves up to 11 cm / a at present.

Hinteres Langtalkar Rock Glacier (HLC): N 46° 59', E 12° 47'; elevation range from 2,455 m to 2,720 m a. s. l. with a length of 900 m and a width of 300 m. HLC is a very active, monomorphic tongue-shaped rock glacier with two rooting zones.

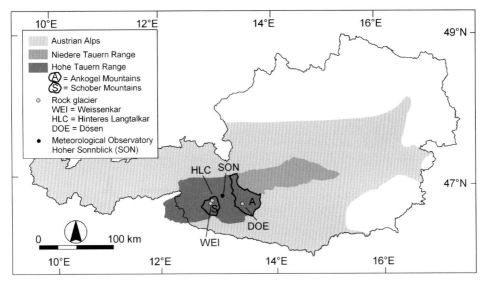

Figure 1: Location of the three rock glaciers Weissenkar (WEI), Hinteres Langtalkar (HLC) and Dösen (DOE) in the Hohe Tauern Range, central Austria. WEI and HLC are located in the sub-unit Schober Mountains, DOE in the Ankogel Mountains. Location of the Meteorological Observatory Hoher Sonnblick (SON) is indicated. The proximity of the three rock glaciers to SON was a big advantage for our analyses because of long-term climatic data series

Distinct changes of the rock glacier surface were detected on aerial photographs from 1997 on (Avian et al. 2005). The movement pattern since 1997 differentiates a slower upper part and a substantially faster lower part with maximum horizontal displacement rates up to 250 cm / a. Therefore, HLC is one of the currently fastest moving rock glaciers in Europe (Delaloye et al. 2008) and possibly in the world underlining the high importance of research continuation.

Dösen Rock Glacier (DOE): N 46° 59', E 13° 17'; elevation range from 2,355 to 2,650 m a. s. l. with a length of 950 m and a width of 250 m. DOE is an active, mono-morphic tongue-shaped rock glacier situated at the end of the glacially shaped, W-E oriented inner Dösen Valley. Displacement measurements (horizontal and vertical) started at this site in 1995 revealing mean surface velocity rates of up 13 to 37 cm / a.

3 Methods

3.1 3D-kinematics of rock glacier

3.1.1 Maintenance of geodetic networks

The geodetic networks of the three rock glaciers consist of stabilized points, mounted either on stable or non-stable, i. e. moving, ground (Fig. 2). For better identification of these points in the field red paint is used to mark and to enumerate them clearly.

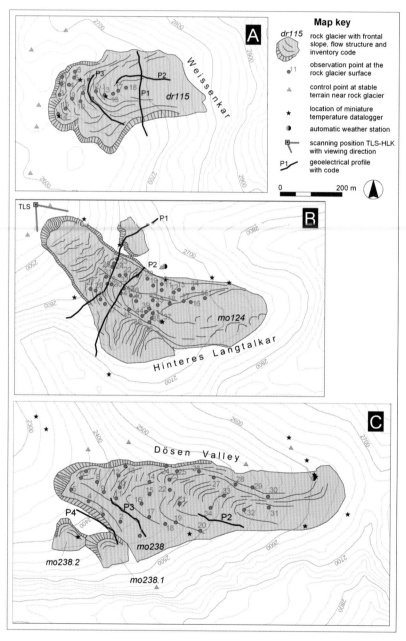

Figure 2: Morphology of the three rock glaciers WEI (A), HLC (B) and DOE (C) with locations of relevant instrumentations and measurement sites. These are locations for geodetic monitoring (observation points at the rock glacier surface and stable control points), locations of miniature temperature data logger (MTD; partly shown – some are outside the depicted maps) and automatic weather stations (AWS), position of the terrestrial laser scanner (TLS) at HLC and locations of the geoelectrical profiles. Locations of very low frequency (VLF) electromagnetic measurements are not indicated. Rock glacier codes according to the inventory by Lieb et al. (2010) described in Kellerer-Pirklbauer et al. (2012)

The maintenance work carried out on an annual basis consisted of (1) check of point existence, (2) check of point stability, (3) exchange of rusty screws, (4) lubrication of the screws, (5) renewal of faded red paint, and (6) renewal / placement of cairns for faster point identification.

3.1.2   Annual geodetic measurements
For the geodetic monitoring (deformation analysis of the rock glacier bodies) appropriate geodetic networks consisting of stable reference points in the surroundings of the rock glacier and observation points on the rock glacier were available. The geodetic measurements from 2010 to 2012 were carried out following the proven scheme of the previous years (Kienast & Kaufmann 2004). The geodetic equipment consisted of a Total Station 1201 of Leica (Fig. 3). Leica circular prisms (geodetic reflectors) were used for point signalling. Data evaluation was done in the office using standard geodetic software (Geosi 6.0). Co-ordinate lists of all observation points of the three rock glaciers were prepared for each of the three epochs. Point accuracies obtained are in the order of ± 1 cm. As a result, annual 3D displacement vectors were computed for all observation points. Products derived therefrom are 2D displacement vectors, annual horizontal flow velocities, strain rates, and other key figures of surface deformation.

3.1.3   Photogrammetric surface deformation measurement
Aerial photogrammetry enables area-wide mapping of surface change. Multi-temporal aerial photographs can be used to detect and measure surface movement. Surface change can be described using 2D or 3D displacement vectors. Knowing the time span between two overflights, change rates, i. e. velocities, can be derived thereof. The successful implementation of a stringent algorithm for computing 3D displacement vectors using multi-temporal digital aerial photographs was demonstrated by Kaufmann & Ladstädter (2003).

Figure 3: Total Station TCRA 1201 of Leica operated in the field

In the framework of the present project, however, the authors applied a modified algorithm which computes 2D displacement vectors based only on high-resolution inter-annual orthophotos taken from virtual globes. Such orthophotos are generally free of charge. However, appropriate orthoimages (high resolution, good radiometry, multi-temporal, known acquisition dates) are scarce. The applicability of the proposed method has been shown for several rock glaciers in the Schober Mountains, Hohe Tauern (Kaufmann 2010; Kaufmann et al. 2012). For reasons of comparison this method was also evaluated at HLC and at Äußeres Hochebenkar rock glacier, Ötztal Alps (Kaufmann 2012). The results obtained for HLC are presented in this report.

3.1.4 Terrestrial laser scanning (TLS)

Usage of TLS on rock glaciers in the European Alps started at the beginning of this millennium (Bauer et al. 2003) and allows acquiring 3 D surface data with high spatial sampling rate. TLS is a time-of-flight system that measures the elapsed time of the laser pulse emitted by a photodiode until it returns to the receiver optics. Maximum range mainly depends on the reflectivity of surface (which is excellent for snow, rock or debris), and atmospheric visibility (best for clear visibility, bad for haze and fog). Since each single measurement consists of a multitude of laser returns, different measurement modes (first return, last return, strongest return) give proper results even during bad weather conditions and on poor surfaces that may otherwise lead to ambiguous measurements like vegetated, moist or roughly structured terrain (Baltsavias et al. 1999).

Figure 4: Terrestrial Laser Scanner RIEGL LMS Z620 in front of the rock glacier HLC at the scanning position HLC_Grat (which was not considered in this study). Codes: (1) Scanning position HLK-TLS1 at the rock glacier front, (2) prominent terrain ridge, (3) transversal furrows/crevasses, (4) location of the automatic weather station. White dotted arrows indicate potential rock glacier nourishment paths at HLC

Long-range TLS (more than 400 m) is of particular interest for measuring high mountain environments as it offers very detailed digital surface models in non-accessible terrain (Bauer et al. 2003). A (theoretical) measuring range of up to 2,000 m (RIEGL LMS Z620) allows hazardous sites to be easily measured from a safe distance. Despite these advantages, TLS has rarely been used in characterizing rock glacier movement (Bodin et al. 2008; Avian et al. 2009)

In 2003 the University of Graz joined JOANNEUM Research Graz in a set of experiments using this new technology for monitoring glaciers and rock glaciers in the Austrian Alps. However, first measurements at the rock glacier HLC and at the Gößnitz Glacier were carried out in summer 2000. First measurements at Pasterze Glacier followed in 2001. All these measurements were carried out using the instrument Riegl LPM–2k (e. g. Bauer et al. 2003; Kellerer-Pirklbauer et al. 2005; Avian et al. 2008, 2009). The recent measurements at HLC (Fig. 4) were carried out within the framework of permAfrost.

From 2009 on a new system (instrument Riegl LMS-Z620) has been used to reduce acquisition time as well aground sampling distance and hence increase point density. Digital Terrain Models (DTM) derived from data of different measurements can be subsequently compared to get a full description of changes in volume, surface dynamics, spatial distribution of shape, or arbitrary profiles on the surface. The filtering and registration of the measured point clouds was conducted in RIEGL RiScan. To avoid misinterpretations all point clouds were matched with the software LS3D (Akca 2010). As reference data set the point cloud measured in 2009 was considered.

3.2   Internal structure of rock glaciers: Geophysics

3.2.1   Very low frequency electromagnetic measurements (VLF)

The VLF method measures the components of the magnetic field of an electromagnetic wave. This electromagnetic wave is transmitted in the majority by military transmitters. The primary purpose of these transmitters with spatial ranges of 4,000 to 5,000 km is to communicate with submarines. The transmitters are installed worldwide. The low frequencies between 15 to 20 kHz had been chosen to get a larger skin depth of the waves. Therefore the depth of penetration of these waves is, compared to waves with higher frequency, deeper.

For geophysical purposes the base wave can be used very well for prospection. The magnetic component of the wave penetrates also deeper into the subsurface. When there are conductive bodies a local change in the amplitude and phase of this magnetic vector is resulting due to the development of secondary fields. In this measurement method the changes of the magnetic component – in this case the tilt angle and the phase shift – are measured by a receiver. Analysing these data by Fraser or Karous-Hjelt filtering (Karous & Hjelt 1977), conductive zones in the subsurface can be detected. We achieve the best resolution with this method when the magnetic vector is normal to the geologic striking. For the current prospection the best suited transmitter in direction and signal strengths is the VLF-transmitter JXZ (Helgeland,

Norway) with a frequency of 17.1 kHz. For the field measurements a VLF-receiver type EM16 from the company Geonics had been used.

3.2.2 Direct current resistivity measurements (DC)

The geoelectric method measures the resistivity of the subsurface by sending electric current into the subsurface using two electrodes and measuring the resulting voltage at two other electrodes. Applying Ohm'schs law and including the geometry of the electrode layout the resistivity of the subsurface and the 3D distribution can be calculated (Kneisel & Hauck 2008). The depth of penetration can be controlled by the distance between the electrodes – large electrode distances give information on deeper layers as the current can penetrate deeper into the subsurface. A shifting in the electrode layout results in lateral resistivity changes along the profile.

Depending on the task 1D, 2D or 3D measurements can be applied. If also temporal changes using repeat measurements are included the measurements are called 4 D-measurements. In the current case 2D-measurements had been applied (measurement system from LGM, Germany). The changes of the electrode positions are automatically controlled by computer. With this automatic measurement system a higher amount of data can be measured in shorter time compared to the manual method. The coverage of the subsurface is therefore increased. In measuring deeper zones by increasing the electrode distances also the shallower layers contribute to the measuring signal, therefore the measured data had to be inverted to separate the individual influences. For further information on the method such as limitations and equivalence we refer to e. g. Koefoed (1979) or Kneisel & Hauck (2008).

3.3 Permafrost and climate monitoring

3.3.1 Climate monitoring

The main devices used in the permafrost monitoring network at the three study areas are (1) automatic energy-balance monitoring stations or automatic weather stations (AWS), (2) miniature temperature data loggers (MTD) for monitoring ground surface and near ground surface temperature, and (3) automatic remote digital cameras (RDC).

The two AWS were installed in 2006 at the rock glaciers DOE and HLC within the ALPCHANGE project (Fig. 5). At both stations, climate data including air temperature, air humidity, wind speed (max, mean), wind direction and global radiation are continuously logged since then although major technical problems caused data gaps. No station exists at the rock glacier WEI, but due to its close distance to HLC (less than 4 km) and the same elevation range, the climate data collected at the rock glacier HLC are presumably also valid for the rock glacier WEI. However, solely air temperature at WEI was measured in the period from 2011 to 2012 for comparison.

3.3.2 Ground temperature monitoring

About 30 miniature temperature data loggers (MTD) for monitoring ground surface, near ground surface and air temperature were installed in 2006 and later at

Figure 5: The AWS at the rock glaciers HLC (A) and DOE (B) installed in 2006. The station at HLC is located on bedrock at 2,655 m a. s. l. in close vicinity to the rock glacier. The station at DOE is located on a large boulder on the rock glacier surface at 2,600 m a. s. l. Photograph viewing directions towards NW (A) and W (B). Note Mt. Großglockner (3,798 m a. s. l.) in the background of (A)

the three rock glacier sites. The used MTDs are either 1-channel data loggers (Geo-Precision, Model M-Log1) monitoring with one temperature sensor or 3-channel data loggers (GeoPrecision, Model M-Log6) monitoring with three sensors at different depths. According to the manufacturer, the used PT1000 temperature sensors have an accuracy of $\pm 0.05\,°C$, a range of -40 to $+100\,°C$ and a calibration drift of $< 0.01\,°C/a$. Generally, the MTDs were funded by the ALPCHANGE project.

Within the following PermaNET and permAfrost projects it was only possible to purchase expendables (such as batteries, tape, etc.) and change or optimize previously established MTD sites at HLC, DOE or WEI. By the end of December 2012 27 MTD sites with 56 temperature sensors for ground temperature monitoring were operating at the three study areas. At 14 sites 1-channel loggers have been used, at further 13 sites 3-channel loggers were installed (Table 1).

In addition to the three main study areas above, MTDs for monitoring ground surface, near ground surface and air temperature were maintained at additional sites in the Hohe Tauern Range (Pasterze Glacier, Fallbichl-Schareck, Hochtor Pass, Hintereggen Valley, and Kögele Cirque). These activities were not funded within permAfrost. Figure 6 depicts the topographical situations and the locations of the MTDs at HLC, WEI and DOE. Furthermore, the situations for the two study areas Kögele Cirque (KC) – located next to HLC – and Pasterze Glacier (PAG) are indicated in this figure as complementary information.

3.3.3 Monitoring of rooting zone processes using remote digital cameras (RDC)

Two automatic remote digital cameras (RDC) were maintained at the two sites HLC and DOE in order to take daily images of the rooting zones of the two rock glaciers. In the rooting zone snow cover conditions (snow cover duration, melt-out date, avalanches) and mass movement events are of particular interest. The RDC system was a self-developed product during the ALPCHANGE project. The RDC consists of a standard digital camera (Nikon Coolpix 4300), a timer control unit (DigiSnap

Table 1: The 27 MTD-sites at the three study areas where ground surface and near surface temperature was monitored during the permAfrost WP4000 project period. Different parameters are indicated for each site. Furthermore, the available data series since 1 June 2010 are listed. Substrate abbreviations: FGM = fine-grained material, CGM = coarse-grained material, BED = bedrock. For locations see Figure 6

Area	MTD-site	Description	Sub-strate	Elevation [m a. s l.]	Aspect [°]	Slope [°]	Sensor depth(s) [cm]	Data series
HLC	HLC-LO-S	debris slope	CGM	2,489	245	32	0	010610-220812
	HLC-MI-S	debris slope	CGM	2,581	268	19	0	010610-220812
	HLC-UP-S	debris slope	CGM	2,696	256	22	0	010610-220812
	HLC-LO-N	rock wall niche with debris	BED	2,485	47	45	0	010610-220812
	HLC-MI-N	debris slope	CGM	2,601	17	28	0	010610-220812
	HLC-UP-N	rock wall niche with debris	BED	2,693	45	52	0	010610-220812
	HLC-RF-S	rock face	BED	2,725	241	75	3, 10, 40	010610-220812
	HLC-RF-N	rock face	BED	2,693	45	85	3, 10, 40	010610-220812
	HLC-RT	flat bedrock site	BED	2,650	252	7	3, 10, 40	010610-220812
	HLC-CO	rock glacier sediments	CGM	2,672	338	8	3, 10, 100	010610-220812
	HLC-SO-S	solifluction lobe	FGM	2,391	253	34	0, 10, 40	010610-220812
	HLC-SO-N	solifluction lobe	FGM	2,407	34	33	0, 10, 40	010610-220812
WEI	WEI-LO	rock glacier sediments	CGM	2,652	238	22	0	010610-210812
	WEI-MI	rock glacier sediments	CGM	2,662	270	3	0, 30, 100	010610-210812
	WEI-UP	rock glacier sediments	CGM	2,688	241	7	0	010610-210812
DOE	DOV-LO-S	debris slope	CGM	2,489	220	22	0	250806-200812
	DOV-MI-S	rock wall niche with debris	BED	2,586	213	19	0	010610-200812
	DOV-UP-S	debris slope	CGM	3,002	166	33	0	010610-300611 & 160811-200812
	DOV-LO-N	debris slope	CGM	2,407	342	22	0	010610-200812
	DOV-MI-N	debris slope	CGM	2,501	239	16	0	010610-200812
	DOV-UP-N	debris slope	CGM	2,626	331	25	0	010610-200812
	DOV-RF-S	rock face	BED	2,628	206	80	3, 10, 32	010610-200812
	DOV-RF-N	rock face	BED	2,638	300	90	3, 10, 40	010610-230711 & 160811-200812
	DOV-RT	flat bedrock site	BED	2,603	255	14	3, 10, 40	010610-200812
	DOV-CO*	rock glacier sediments	CGM	2,606	257	5	100, 200, 300	010610-200812
	DOV-FI	slope with veg-etation	FGM	2,644	213	28	0, 3, 10, 30, 70, 100	010610-200812
	DOV-SO	solifluction lobe	FGM	2,578	116	18	3,10,70	No data

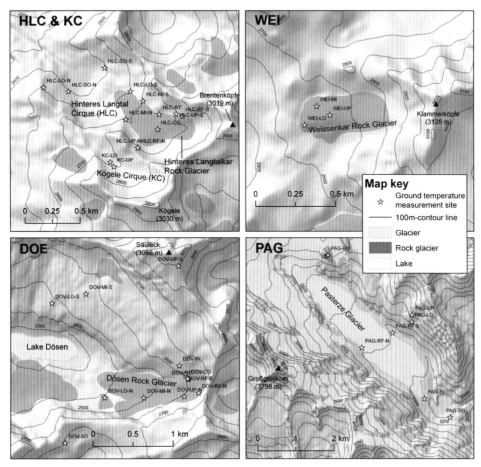

Figure 6: Detailed maps of the three main study areas HLC, WEI and DOE as well as of two additional study areas at KC (Kögele Cirque) and PAG (Pasterze Glacier) with the locations of ground temperature measurement sites using miniature temperature datalogger (MTD). For details and description of the MTD sites at the three main study areas refer to Table 1

2000) and a weatherproof case all mounted on a 1.5 m high steel pole and power supplied by solar panel connected to a storage battery.

The RDC system at HLC was installed on the ridge between HLC and KC at 2,770 m a.s.l. in mid September 2006 (Fig. 7). The one at DOE was installed in early September 2006 at a small rock hammock overlooking the rock glacier at an elevation of 2,630 m a.s.l.. During the permAfrost project period, RDC data at HLC were collected continuously during the period 19.08.2010 to 15.07.2012 (697 daily images). At DOE the situation is quite different with shorter data availability during the two periods 17.08.2010 to 9.12.2010 (116 images) and 16.08.2011 to 14.2.2012 (183 images). Reason for the failure of the cameras might be technical problems with electronic parts caused by the harsh climatic conditions.

Figure 7: Two photographs of the RDC system at HLC taking daily images from the rooting zone of the rock glacier

4   Results

4.1   3D-kinematics of rock glaciers

4.1.1   Annual geodetic measurements
All dates of geodetic measurement for the three rock glaciers are listed in Table 2. The reference epoch for comparison is 2009. The table clearly shows that the measurements were carried out almost at the same dates as during the other measurement years with a temporal deviation of only a few days.

In the following, the main results obtained for the three rock glaciers will be presented briefly. Maximum flow velocities obtained at all three rock glaciers are presented in Table 3. The maximum flow velocity is a highly characteristic parameter for rock glacier flow and it can be beneficially used in rock glacier movement analysis (Kellerer-Pirklbauer & Kaufmann 2012). Selected results of horizontal displacement are depicted in Figures 8 to 10 (see also Kaufmann 2013a–c).

Flow rates at HLC were exceptionally high throughout the whole monitoring period 1999 to 2012 (Fig. 11). Maximum values were measured for the last measurement year 2011 to 2012. These high flow rates can be most probably explained by the downwasting of large rock masses at the frontal slope. The rock glacier is moving over a terrain ridge into steeper terrain. This geomorphic process is associated with a marked disintegration of the rock glacier surface and the development of surface ruptures (see Fig. 13).

Table 2: Dates of geodetic measurements at DOE, HLC and WEI

Rock glacier	2009	2010	2011	2012
DOE	August, 18	August, 17	August, 16	August, 14
HLC	August, 21	August, 21	August, 20	August, 18
WEI	August, 22	August, 20	August, 19	August, 17

Table 3: Maximum flow velocities measured at DOE, HLC and WEI. Observation points are listed within brackets

Rock glacier	2008–2009	2009–2010	2010–2011	2011–2012
DOE	39.7 cm / a (15)	41.7 cm / a (15)	43.1 cm / a (15)	44.5 cm / a (15)
HLC	1.75 m / a (23)	2.43 m / a (23)	2.94 m / a (23)	3.41 m / a (25)
WEI	7.0 cm / a (14)	10.1 cm / a (14)	10.7 cm / a (14)	13.1 cm / a (16)

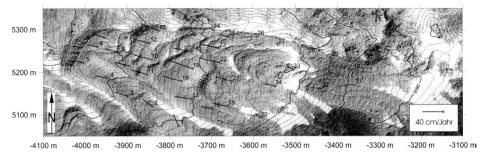

Figure 8: 2D flow vectors at DOE for the time period 2011 to 2012. A maximum flow velocity of 44.5 cm/a was measured at point 15. The total number of observation points on the rock glacier is 109: 34 stabilized points (see this Figure and Fig. 2) and 75 paint-marked points

Figure 9: 2D flow vectors at HLC for the time period 2011 to 2012. A maximum flow velocity of 3.41 m/a was measured at point 25. The total number of observation points on the rock glacier is 38 (see Fig. 2)

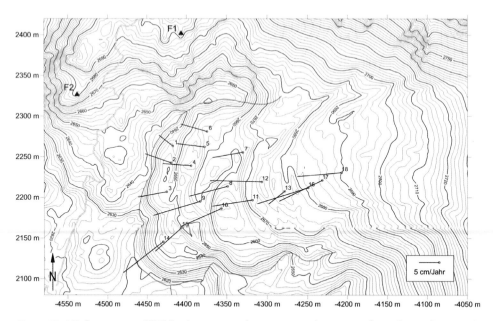

Figure 10: 2 D flow vectors at WEI for the time period 2011–2012. A maximum flow velocity of 13.1 cm/a was measured at point 16. The total number of observation points on the rock glacier is 18 (see Fig. 2)

Figure 12 shows the temporal change of the mean annual flow velocity for each of the three rock glaciers for the available geodetic data sets. Most interestingly the three curves shown suggest a high correlation of the flow velocities measured. Acceleration and deceleration, respectively, observed at the three rock glaciers are to a high degree synchronous. A possible explanation for this phenomenon give Kellerer-Pirklbauer & Kaufmann (2012). We conclude that climate parameters control to a certain extent rock glacier flow and particularly relative changes in flow velocities.

4.1.2   Photogrammetric surface deformation measurement
High-resolution orthoimages of Google Maps and Microsoft Bing Maps covering HLC date from 18.9.2002 and 21.9.2006 (Fig. 13). Displacement vectors were computed following the procedure of Kaufmann (2010). Based on this information a colour-coded velocity map was generated (Fig. 14). A detailed accuracy analysis of the results obtained and further technical information on the analyses is given in Kaufmann (2012).

4.1.3   Surface deformation measurements from Terrestrial Laserscanning
TLS measurements at the rock glacier HLK were carried out on 6.8.2009, 15.9.2010, 25.8.2011, and 1.8.2012. The accuracies of registration of each single point cloud are in the order of ± 2.0 cm considering at least five tie points. In this analysis simple elevation differences of DTMs of each epoch were calculated. Therefore all results do not represent displacement vectors in a strict sense than a vertical change of a pixel

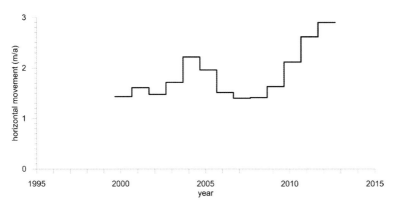

Figure 11: Mean annual horizontal movement of the 6 marked points (23–25, 27, 28, 31) for the time period 1999 to 2012. All 6 points are located at the lower, faster part of the rock glacier HLC

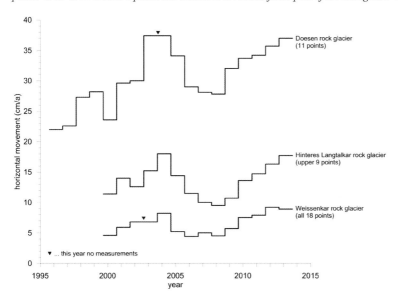

Figure 12: Comparison of the mean annual flow velocities obtained at DOE, HLC and WEI. 11 points at DOE: 10–17, and 21–23; 9 points at HLC: 10–17 and 37 (all points are from the upper part of the rock glacier which is substantially slower compared to the lower part; see Fig. 11); all 18 points at WEI

in a positive (surface lifting) or negative (surface decline) direction which is below termed as vertical surface dynamics.

Table 4 presents mean vertical surface elevation changes for different areas at the front of the rock glacier HLC based on the TLS campaigns. Mean vertical surface elevation changes over the entire rock glacier tongue (extent see Fig. 15) are in the range of 39 to 45 cm / a with a peak at the epoch 2010 / 11. For all three epochs, four areas with similar patterns of vertical surface dynamics can be differentiated. The

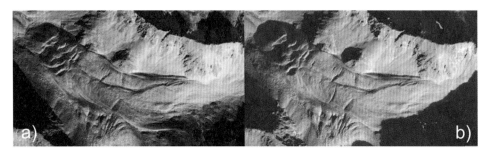

Figure 13: This figure shows a temporally mixed stereogram of HLC at 80 cm ground sampling distance: a) Google Maps, epoch 2002, b) Microsoft Bing Maps, epoch as at 2006.

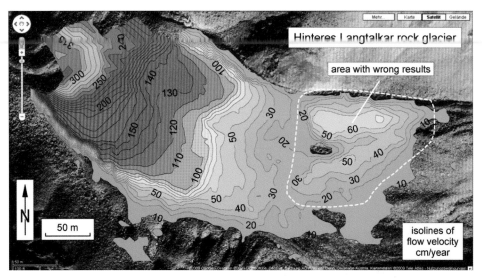

Figure 14: Isolines of mean annual horizontal flow velocity of HLK derived from image data of Google Maps (epoch 2002) and Microsoft Bing Maps (epoch 2006)

four similar areas are exemplarily delineated for the epoch 2009 to 2010 shown in Figure 15.

1. The lowest part of the front of the rock glacier HLC is quite stable in terms of small vertical surface elevation changes. This part is characterized by accumulated loose debris transported down from the adjacent moving part uphill (Table 4 / 1).

Table 4: Mean vertical surface elevation changes (→ dynamics) from TLS at the entire front lobe and the four distinct parts of HLC from TLS

Rock glacier HLC	2009–2010	2010–2011	2011–2012
Mean	0.39 m / a	0.47 m / a	0.39 m / a
Area (1)	0.08 m / a	0.11 m / a	0.13 m / a
Area (2)	0.64 m / a	0.65 m / a	0.57 m / a
Area (3)	1.12 m / a	0.80 m / a	0.80 m / a
Area (4)	0.98 m / a	0.97 m / a	0.79 m / a

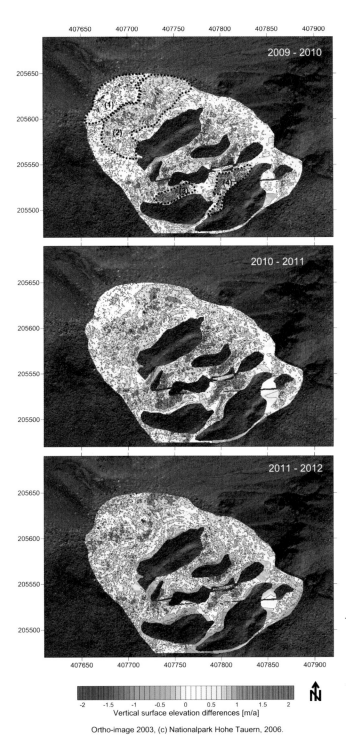

Figure 15: Mean annual vertical surface elevation differences of rock glacier HLC at the epochs 2009 to 2010, 2010 to 2011, and 2011 to 2012. In epoch 2009 to 2010 the delineation of distinct patterns of mean annual vertical surface elevation differences are shown exemplarily and are indicated with the codes (1) to (4)

2. Adjacent uphill, a large area with high vertical movement rates (> 0.57 m / a at all epochs) can be detected. This vertical surface lifting can be equalized – in terms of geomorphological process interpretation – with an intense downslope movement (Table 4 / 2).
3. The upper part of the rock glacier front can be differentiated in a left and central part with high positive vertical movement rates (> 0.80 m / a at all epochs, Table 4 / 3 as well as 4).

4.2   Internal structure of rock glaciers based on geophysical measurements

Geophysical field work at the three rock glaciers was carried out during the period 17.8.2011 to 6.9.2011. At HLC two geoelectrical profiles and 148 VLF points were measured. At WEI three geoelectrical profiles and 126 VLF points were measured. Finally, at DOE four geoelectrical profiles (although only three successfully) and 122 VLF points were measured. The locations of the profiles are indicated in Figure 2. Results of VLF and from geoelectrical profiles do coincide in terms of high and low resistivity values but cannot be compared directly. Although results of VLF seem to be coarse in terms of resolution, results seem to distinguish areas without permafrost from ice rich areas.

4.2.1   Indicators for internal structure at HLC

Due to very unfavorable terrain conditions and time consuming campaigns only areas between the geoelectric sections were measured with 148 VLF-points at the rock glacier HLC. VLF-results show a shift of resistivity from NE to SW with a low resistivity anomaly in NW and a high resistivity anomaly in SE and therefore a change in the inner structure of the rock glacier (Fig. 16, left).

The two geoelectrical profiles at HLC (Fig. 2 for locations) are briefly described. Profile HLC-P1 (50 electrodes, length 490 m) started from bedrock (BR, "profilemeter" PM1–60 and from PM460 on, overlain by a shallow debris cover), followed by a slope section of blocky debris (BD, depth 10 m, PM60–PM140). Under this scree slope low values of resistivity were measured which can be interpreted as jointed bedrock allowing groundwater flow and hence reduced resistivity values. Adjacent (PM140–PM220) measurements suggest a rather dry bedrock area. The section between PM215 and PM460 comprises ice rich areas which can be indicated as the main rock glacier body. Resistivity values suggest two homogeneous units whereas the section with significantly lower values covers PM215–PM280 with depths up to 20 m as well as the section with significantly higher values from PM280–PM365 with depths up to 15 m. Between PM200 and PM380 a layer of very low resistivity values was measured in depths of > 40 m.

Profile HLC-P2 (30 electrodes, length 290 m) is characterized by similar measurement values as in Profile HLC-P1, although areas with very high resistivity – compared to HLC-P1 – were measured too (Fig. 17). The beginning of the profile was set on bedrock as at HLC-P1 (until PM60), which is also overlain by a shallow debris

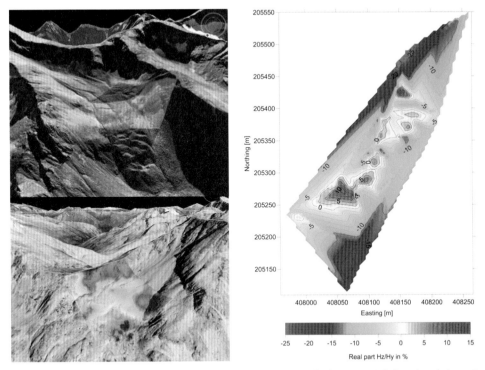

Figure 16: Left – 3D visualization of interpolated VLF-data at rock glacier HLC (left-top) and the rock glacier WEI (left-bottom). Right – Isolines of interpolated VLF-data for HLC. Legend is valid for all three sub-figures

cover (5 m). The rock glacier body itself begins at PM60 with a significant increase of resistivity values. From PM70 very high resistivities with depths to 20 m were measured until PM190 and down to 10 m until PM220.

4.2.2 Indicators for internal structure at WEI

At rock glacier WEI three geoelectric profiles (Fig. 2) and 126 VLF points were measured. Generally it can be stated that all resistivity measurements are one magnitude higher than measurements on rock glacier HLC.

Profile WEI-P1 (30 electrodes, length 290 m, alignment: N–S): WEI-P1 transects the rock glacier transversally. High resistivities are detectable beginning with PM20 consistently until PM145, thicknesses vary from 10 m at the N of the rock glacier to 15 m at left margin in the S of WEI. Thickness of the active layer is about 4–5 m.

Profile WEI-P2 (20 electrodes, length 190 m, alignment: W–E, intersected WEI-P1 orthogonal at PM100): The results from longitudinal profile WEI-P2 correspond well with WEI-P1 in the transitions-zone. At PM45 resistivity values increase significantly indicating an ice-rich area with a thickness of 20 m until PM150.

Profile WEI-P3 (20 of 23 electrodes usable, length 220 m, at rock glacier front): This section shows evidences of ice-rich substrate beginning with PM30 until PM110

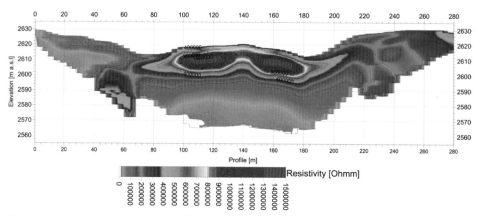

Figure 17: Results at the geoelectrical profile P2 and the rock glacier HIC

under active layer of 4 to 6 m (Fig. 18). At this part of the rock glacier ice rich substrate – hence permafrost – seems to occur as a lens. This seems feasible considering a radiation favoring location (W-facing slope), the relative low elevation and the relative low movement rates as revealed by the annual geodetic campaigns.

4.2.3   Indicators for internal structure at DOE

At DOE four geoelectrical profiles were measured originally although only three successfully (Fig. 2) because of problems with conductivity occurring at profile DOE-P1 during the field campaign. The results from the remaining three profiles are briefly described.

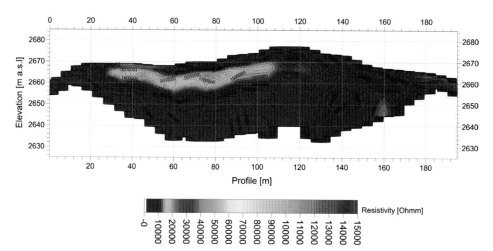

Figure 18: Results at the geoelectrical profile P3 and the rock glacier WEI

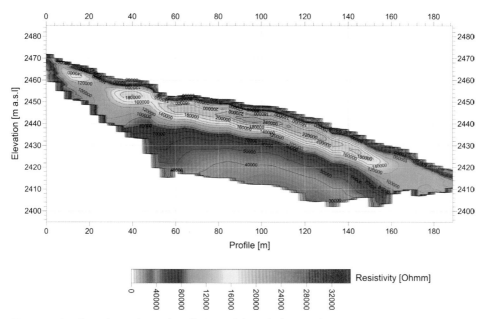

Figure 19: Results at the geoelectrical profile P3 and the rock glacier DOE

Profile DOE-P2: The profile transects the upper part of the rock glacier longitudinally / downslope (30 electrodes, 5 m distance = length 145 m, alignment: E–W). From PM40 on high resistivities at a depth of 5 m indicate ice-rich substrate. This presumably solid permafrost body of at least 10 to 15 m thickness – in horizontal as well as in vertical extent – distinctly ends at PM115.

Profile DOE-P3 (Fig. 19): The profile follows a ridge at the orographic left part of the lower rock glacier area (34 electrodes, 6 m distance = length 198 m, alignment: SE–NW). At this longitudinal profile evidences of a permafrost body start at PM8 with high resistivities in a depth of 4 m and a thickness of 4 to 5 m to PM20 followed by low resistivities until PM35.

From PM35 on high resistivities in a depth of 3 to 5 m with a thickness of 12 to 15 m, continuing until PM160. The near surface sediments consisting of large blocks and fine grained substrate show thicknesses of 4 to 5 m, the thickness of ice-rich areas (permafrost body) shows continuous thicknesses of 12 to 15 m.

Profile DOE-P4: The profile follows the orographic left (southern) margin of the rock glacier body. (40 electrodes, 5 m distance = length 195 m, alignment: SE–NW). At this profile only marginal evidences of permafrost with thicknesses of ca. 8 m between PM35 and PM75 were detected (lowest elevation: 2,410 m a. s. l.). Due to the setting on the left margin of the rock glacier it clearly shows a transition of a permafrost area to a permafrost-free area.

4.3   Permafrost and climate monitoring

4.3.1   Meteorological data at DOE and HLC

Table 5 lists the collected meteorological data with data gaps and periods of field campaigns between 2010 and 2012. The data gaps are related to the harsh environmental conditions and animals damaging sensors or cables. When possible, sensors were repaired or changed during the field campaigns. Some details for example: One anchoring wire rope at the AWS HLC was destroyed presumably by a lightning stroke in 2009 to 2010 and repaired in 2010 and further improved in 2011. At the DOE AWS site, the sensor collecting relative humidity and air temperature was destroyed during the measurement year 2010 to 2011. Therefore, a new combined relative humidity-air temperature sensor (EE08, EE Electronics, Austria) was installed in 2011.

Data from the two AWS sites were for instance used in the analyses for the publication regarding the relationship between rock glacier velocity and climate parameters thereby analysing the three different rock glaciers (Kellerer-Pirklbauer & Kaufmann 2012). In this work climate data were combined with ground temperature data and rock glacier velocity data. By correlating rock glacier velocity with different climatic parameters (air temperature, snow depth, ground surface temperature and ground temperature at 1 m depth and derivatives of them) it was shown that the relationships are complex and only in few cases statistically significant. Most of the correlating pairs of variables support the observation that warmer air temperatures, warmer ground surface temperatures, and warmer ground temperatures at 1 m depth (or derivatives of the three parameters) favour faster rock glacier movement and therefore rock glacier acceleration. The most striking problem during the analysis were short time series in particular of ground temperature (surface and at one meter depth). For details on these analyses refer to Kellerer-Pirklbauer & Kaufmann (2012).

Table 5: Data series collected from the two automatic weather stations (AWS) at the two study areas DOE and HLC during the permAfrost-WP4000 project period starting on 1 June 2010. Periods of field campaigns to the study areas DOE and HLC including WEI are indicated in the last row

Station	Data series	Comment	Field campaigns
AWS-DOE	010610-150511	Partly data available; no data for air humidity	16.–18.08.2010
	150511-160811	Partly data available; no data for air humidity and air temperature	15.–17.08.2011 17.–18.10.2011
	160811-090212	Data series completely available, no problems	19.–21.08.2012 24.–25.09.2012
	100212-080712	Partly data available; no data for global radiation	
	080712-200812	Partly data available; no data for global radiation and maximum wind speed	
AWS-HLC	010610-150910	No data available; problem with lightning stroke effects	18.–21.08.2010
	160910-060112	Data series completely available, no problems	15.–16.09.2010 17.–20.08.2011
	060112-220812	Partly data; sensor for air temperature and humidity malfunctioned due to cable destruction	21.–25.08.2012

4.3.2   Ground temperature data at HLC, WEI and DOE

In general, the MTDs worked quite well at all three study sites and no major problems were faced during the field campaigns. Results presented in Kellerer-Pirklbauer (2013) clearly revealed ground surface warming at most MTD sites since 2006. Over the last years warming was stronger in the western part of the national park (study areas HLC and WEI) in relation to the eastern part (DOE). This warming generally influences all elevations above ca. 2,000 m a.s.l., all slope inclinations, all aspects and seems to be unrelated to winter snow cover conditions at the MTD sites as indicated by the analyses. Figure 20 shows the trend results for the 27 MTD sites in our three main study areas based on all available data for each site using linear regression. Furthermore, Figure 21 shows two examples of temperature evolution (based on mean daily data) since the beginning of the measurements: one example reveals a warming trend (HLC-UP-S), the second example reveals no trend at all (DOV-UP-S). Both sites are permafrost-sites as indicated by the base temperature of the winter snow cover (BTS) in late winter at HLC-UP-S (Haeberli 1973) for HLC-UP-S or the low mean annual ground surface temperature for DOV-UP-S. For details and other analyses based on the MTD-data refer to Kellerer-Pirklbauer (2013).

4.3.3   Optical monitoring based on the RDC data at HLC and DOE

The RDC system at HLC has been running more stable compared to DOE with shorter data gaps. Particularly the good data quality at HLC allowed a detailed analysis which were analysed primarily within the framework of a master thesis (Rieckh 2011) and presented at different conferences (Kellerer-Pirklbauer et al. 2010; Rieckh et al. 2011). Figure 22 exemplarily shows snow cover duration maps for the hydrological years 2006 to 2007 and 2007 to 2008 in the north-eastern rooting zone of the

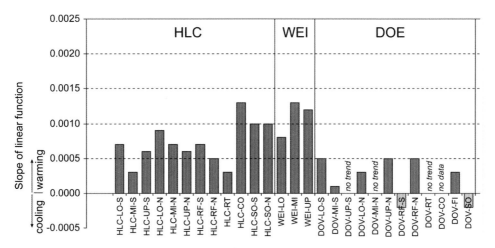

Figure 20: Slope of linear function for ground surface temperature using linear regression for all MTD-sites at the three study areas WEI, HLC and DOE. The calculated trends are based on all available daily mean data per site. In most cases data series start in 2006

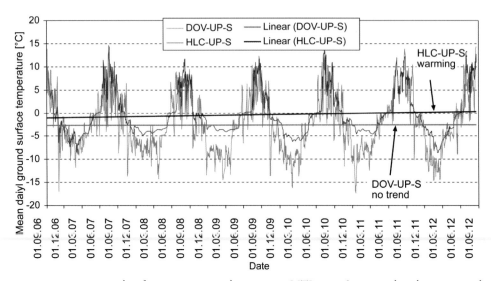

Figure 21: Mean ground surface temperature evolution at two MTD sites. One example indicates no trend and hence neither warming nor cooling of the surface at all (DOV-UP-S), a second example indicates a clear warming trend (HLC-UP-S)

rock glacier HLC. Note the distinct differences in the duration but the general similarity in the pattern. The duration difference is related to the snow-poor conditions during the first year and rather normal snow conditions during the second year (as can be judged by data comparison with snow data from the neighbouring meteorological observatory at Mt. Hoher Sonnblick).

5   Synthesis of results

5.1   Synthesis for WEI

Rock glacier WEI is the slowest moving one of the three studied rock glaciers. This rock glacier is also different to the other two rock glaciers because of the radiation exposed site, the relatively warm ground surface temperatures measured at three sites at the rock glacier surface and the results for the geophysical measurements indicating a discontinuous permafrost body in the rock glacier sediments possibly indicative for a tendency of this rock glacier to turn to climatic inactive (Barsch 1996).

At two of the three MTD sites mean ground surface temperatures are positive during the period 2006 to 2012. In addition to that, all three sites show a clear warming trend during this period (Fig. 21). Furthermore, geoelectrical profiles at the central part of the rock glacier WEI show clear evidence of ice rich substrate under an active layer of 4 to 5 m. The thickness of the permafrost body is assumed to be ca. 15 to 20 m, the transversal extent along the profile WEI1 is expected with ca. 120 m. Therefore ice is very likely along most of the rock glacier. Permafrost in the lowest

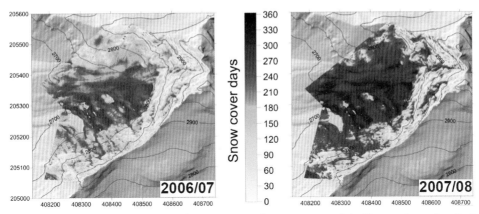

Figure 22: Snow cover duration maps for two years at the rooting zone of the Hinteres Langtalkar Rock Glacier based on daily optical data from the RDC system installed above the area of interest. For analytical approach and technical details see Rieckh (2011)

part of the rock glacier WEI can be expected as marginal and if so only expressed by isolated lenses. This argument is comprehensible considering a radiation favoring location, the rather low elevation and the rather low movement rates as revealed by the annual geodetic campaigns.

5.2 Synthesis for HLC

Rock glacier HLC is very special because of the disintegration processes which started in the mid 1990s causing tearing-apart surface structures. Transversal assemblages of furrows and ridges with very high rate (for rock glaciers at least) at the frontal part and substantially lower rates at the upper-most part determine the appearance of the rock glacier particularly. This peculiarity makes this rock glacier of special interested leading to permafrost-related research at this rock glacier since the 1990s (see Kellerer-Pirklbauer & Kaufmann 2012 for complete references).

The lower part of the rock glacier HLC is characterized by different areas of the vertical surface elevation differences between the epochs with distinct patterns. Disintegration and extensive mass wasting especially at the very front of the rock glacier still proceeds making in situ field work also very difficult and dangerous. Mean values of vertical surface elevation changes at the frontal part are in the range of 40 to 45 cm / a, which is significantly higher than values of 30 to 35 cm / a in the period 2005 to 2008 (Avian et al. 2009). This also coincides with 3D deformation measurements at the middle (Fig. 12) and upper parts (Fig. 14) of the rock glacier HLC. Summarizing it can be stated that different measurements show that the 3D kinematics of the rock glacier increase since the epoch 2007 to 2008 after a period of decrease since a first peak at the epoch 2004 to 2005. Furthermore, ground surface temperatures clearly indicate a warming trend since 2006 which is in agreement with the possible effects of ground warming explained above.

DC-resistivity measurements have been carried out just above the upper part of the TLS measurements and in between the lowest parts of geodetic surveys. Results show that permafrost is very likely in a depth of 3 to 5 m with thicknesses of about 20 m with an extent of 120 m along the profile. As mentioned above displacement rates increase from the epoch 2008 / 2009 to 2011 / 2012. All measurements at the rock glacier HLC show prolonging very high movement rates at the front of the rock glacier and thus of strain related processes such as disintegration and frequent mass wasting at the frontal zone.

5.3   Synthesis for DOE

The rock glacier DOE shows surface movement rates which numerically are in between the rates of WEI und HLC (lower part). In contrast to HLC this rock glacier does not have any crevasse-like structures due to strain or mass-wasting. DOE is rather homogenous in its movement pattern (Fig. 8) and can be regarded as a "standard" rock glacier at that size. Interestingly, mean ground temperatures at the MTD sites are cooler compared to the MTD sites at HLC and WEI. Furthermore, during the last six years no clear trend in ground surface warming or cooling for this rock glacier was detected.

Geoelectrical profiles at the central part of the rock glacier DOE show clear evidence of ice-rich substrate under an active layer of 4 to 5 m. The thickness of the permafrost body is assumed to be appr. 10 to 15 m evenly along the entire longitudinal profile DOE 3 at the central part of the rock glacier. Profiles at the margin of the rock glacier (P2, P4) show a marginal continuation of ice-rich substrate to the adjacent N-facing slopes which proofs the assumption of ice lenses within these scree slopes.

6   Conclusion and outlook

The annual displacement vectors obtained are highly congruent at each of the three rock glaciers. Flow / creep directions are mostly consistent throughout a longer period of time, i. e. 2 to 3 years. Furthermore and most importantly, the geodetic measurements 2010 to 2012 confirmed once more our working hypothesis that the annual changes of flow velocity at the three rock glaciers are highly correlated with one another due to common, external forces, such as air temperature (see a more detailed analysis in Kellerer-Pirklbauer & Kaufmann 2012). At Hinteres Langtalkar rock glacier maximum flow velocities of up to 3.41 m / a were measured, at Dösen 44.5 cm / a, and at Weissenkar rock glacier 13.1 cm / a, respectively.

In order to better study the influence of climate change on rock glacier movement it is necessary to continue the long-term geometric monitoring at least at this time scale, i. e. a one year interval. However, to better address geomorphic process understanding at rock glaciers the authors propose the installation of a continuous geometric monitoring system which is capable of providing daily flow / creep rates.

This could be accomplished by a wireless network of Real-time Kinematic (RTK) low-cost Global Navigation Satellite System (GNSS) receivers. Such a system should then be augmented by other important sensors, such as temperature loggers or hydraulic pressure gauges.

Furthermore it is evidently important to continue and extent the long-term thermal monitoring at these three rock glacier sites, but also other "key-sites" in the Austrian Alps. These sites should be well distributed in different permafrost environments from Western to Central and Eastern Austria. Furthermore, this distribution should consider permafrost sites in the north (Northern Calcareous Alps), in the centre (Central Alps such as e. g. Hohe Tauern Range) as well as the south (Southern Calcareous Alps) of the Austrian Alps. As indicated by our measurements, temperature trends at MTD sites indicate warming in the Schober and Glockner Mountains whereas in the Ankogel Mountains to the east, no clear trends were revealed indicating more stable (permafrost-) conditions in this mountain group. However, obviously longer time series of ground temperature would be certainly helpful to clearer detect such signals.

Geophysical surveys delivered satisfying results although – as methodologically had to be expected – absolute values were not comparable between each rock glacier. Generally differences in resistivity-measurements within the permafrost body can be explained by changes in temperature, thickness, fine grained fraction, and/or stress. Furthermore usage of very low frequency (VLF) measurements showed its feasibility for determining ice/no-ice areas within permafrost areas with very promising results for overview purposes. Improved methodology and implementation of VLF is of great interest for upcoming analyses in high mountain permafrost areas.

Summarizing, this research report clearly illustrates that the combination of the different methods enabled the generation of valuable research results. However, data processing is partly still ongoing and more research results will be published in the future. Furthermore, usage of further methods and instrumentations e. g. continuous movement monitoring, boreholes in permafrost, and other geophysical methods would be very helpful to improve the understanding of rock glaciers.

7   Acknowledgements

This research project was carried out within the project "permAfrost – Austrian Permafrost Research Initiative" funded by the Austrian Academy of Sciences. Furthermore, relevant data were collected earlier within the two projects "ALPCHANGE – Climate Change and Impacts in Southern Austrian Alpine Regions" financed by the Austrian Science Fund (FWF) through project no. FWF P18304-N10 as well as "PermaNET – Permafrost long-term monitoring network". PermaNET is part of the European Territorial Cooperation and co-funded by the European Regional Development Fund (ERDF) in the scope of the Alpine Space Programme. Furthermore, we are very thankful to Dr. Devrim Akca for providing the software LS3D (ETH-Zurich).

Unfortunately, Dr. Erich Niesner deceased during the project period. He was a sincere colleague with splendid ideas allowing fruitful discussions. In order to honor his valuable work during the years but also to honor him as a person, we dedicate this project report in memory of Erich.

8   References

Akca, D. 2010: Co-registration of surfaces by 3D Least Squares matching. *Photogrammetric Engineering and Remote Sensing* 76, 3: 307–318.

Avian, M., G.K. Lieb & V. Kaufmann 2005: Recent and Holocene dynamics of a rock glacier system – the example of Langtalkar (Central Alps, Austria). *Norsk Geografisk Tidsskrift – Norwegian Journal of Geography* 59: 1–8.

Avian, M., A. Kellerer-Pirklbauer & A. Bauer 2008: Remote Sensing Data for Monitoring Periglacial Processes in Permafrost Areas: Terrestrial Laser Scanning at the Rock Glacier Hinteres Langtalkar, Austria. In: Kane, D.L. & K.M. Hinkel (eds.): *Proceedings of the Ninth International Conference on Permafrost,* Fairbanks, Alaska, 29 June–3 July 2008. Vol. 1: 77–82.

Avian, M., A. Kellerer-Pirklbauer & A. Bauer 2009: LiDAR for monitoring mass movements in permafrost environments at the cirque Hinteres Langtal, Austria, between 2000 and 2008. *Natural Hazards and Earth System Sciences* 9: 1087–1094. doi:10.5194/nhess-9-1087-2009.

Avian, M. & A. Kellerer-Pirklbauer 2012: Modelling of potential permafrost distribution during the Younger Dryas, the Little Ice Age and at present in the Reisseck Mountains, Hohe Tauern Range, Austria. *Austrian Journal of Earth Sciences* 105, 1: 140–153.

Baltsavias, E.P. 1999: A comparison between photogrammetry and laserscanning. *ISPRS Journal of Photogrammetry and Remote Sensing* 54, 2-3: 83–94.

Barsch, D. 1996: *Rockglaciers. Indicators for the Present and Former Geoecology in High Mountain Environments.* Springer Series in Physical Environment 16. Berlin.

Bauer, A., G. Paar & V. Kaufmann 2003: Terrestrial laser scanning for rock glacier monitoring. In: Phillips M., S.M. Springman & L.U. Arenson (eds.): *Permafrost: Proceedings of the 8th International Conference on Permafrost,* Zurich, Switzerland, 21–25 July 2003: 55–60.

Berthling, I. 2011: Beyond confusion: rock glaciers as cryo-conditioned landforms. *Geomorphology* 131: 98–106.

Bodin, X. & P. Schoeneich 2008: High-resolution DEM extraction from terrestrial LIDAR topometry and surface kinematics of the creeping alpine permafrost: The Laurichard rock glacier case study (Southern French Alps). In: Kane, D.L. & K.M. Hinkel (eds.): *Proceedings of the Ninth International Conference on Permafrost,* Fairbanks, Alaska, 29 June–3 July 2008. Vol. 1: 137–142.

Boeckli, L., A. Brenning, S. Gruber & J. Noetzli 2012: A statistical permafrost distribution model for the European Alps. *The Cryosphere* 6: 125–140. doi:10.5194/tc-6-125-2012.

Delaloye, R., E. Perruchoud, M. Avian, V. Kaufmann, X. Bodin, H. Hausmann, A. Ikeda, A. Kääb, A. Kellerer-Pirklbauer, K. Krainer, Ch. Lambiel, D. Mihajlovic, B.Staub, I. Roer & E. Thibert 2008: Recent interannual variations of rock glacier creep in the European Alps. In: Kane, D.L. & K.M. Hinkel (eds.): *Proceedings of the Ninth International Conference on Permafrost,* Fairbanks, Alaska, 29 June–3 July 2008. Vol. 1: 343–348.

French, H.M. 1996: *The Periglacial Environment.* Harlow.

Haeberli, W. 1973. Die Basis-Temperatur der winterlichen Schneedecke als möglicher Indikator für die Verbreitung von Permafrost in den Alpen. *Zeitschrift für Gletscherkunde und Glazialgeologie 9,* 1–2: 221–227.

Haeberli, W., C. Guodong, A.P. Gorbunov & S. Harris 1993: Mountain permafrost and climate change. *Permafrost and Periglacial Processes* 4: 165–174.

Haeberli, W., B. Hallet, L. Arenson, R. Elocin, O. Humlum, A. Kääb, V. Kaufmann, B. Ladanyi, N. Matsuoka, S. Springman & D. Vonder Muehll 2006: Permafrost creep and rock glacier dynamics. *Permafrost and Periglacial Processes* 17: 189–214. doi:10.1002/ppp.561.

Karous, M. & S.-E. Hjelt 1977: Determination of apparent current density from VLF measurements. Contribution 89, Department of Geophysics, University of Oulu. http://www.cc.oulu.fi/~mpi/Softat/pdfs/KarousHjelt_1977.pdf (accessed: 3.3.2014).

Kaufmann, V. 2010: Measurement of surface flow velocity of active rock glaciers using orthophotos of virtual globes. *Geographia Technica*, Special Issue: 68–81.

Kaufmann, V. 2012: Detection and quantification of rock glacier creep using high-resolution orthoimages of virtual globes. *The International Archives of the Photogrammetry, Remote Sensing and Spatial Information Sciences* XXXIX-B5: 517–522. doi:10.5194/isprsarchives-XXXIX-B5-517-2012.

Kaufmann, V. & R. Ladstädter 2003: Quantitative analysis of rock glacier creep by means of digital photogrammetry using multi-temporal aerial photographs: two case studies in the Austrian Alps. In: Phillips, M., S.M. Springman & L.U. Arenson (eds.): *Proceedings of the 8th International Conference on Permafrost, Zürich, Switzerland, 21–25 July 2003*: 525–530.

Kaufmann, V., J.O. Filwarny, K. Wisiol, G. Kienast, V. Schuster, S. Reimond & R. Wilfinger 2012: Leibnitzkopf Rock Glacier (Austrian Alps): Detection of a fast Moving Rock Glacier and Subsequent Meassurement of its Flow Velocity. In: *Extended Abstracts of the 10th International Conference on Permafrost* 4, 1. Salekhard, Russia, 25–29 June 2012: 249–250.

Kaufmann, V. 2013a: Doesen Rock Glacier (Carinthia, Austria). http://www.geoimaging.tugraz.at/viktor.kaufmann/Doesen_Rock_Glacier/index.html (accessed: 3.3.2014).

Kaufmann, V. 2013b: Hinteres Langtalkar Rock Glacier (Carinthia, Austria). http://www.geoimaging.tugraz.at/viktor.kaufmann/Hinteres_Langtalkar_Rock_Glacier.html (accessed: 3.32014).

Kaufmann, V. 2013c: Weissenkar Rock Glacier (Carinthia, Austria). http://www.geoimaging.tugraz.at/viktor.kaufmann/Weissenkar_Rock_Glacier.html (accessed: 3.32014).

Kellerer-Pirklbauer, A. 2008: *Aspects of glacial, paraglacial and periglacial processes and landforms of the Tauern Range, Austria*. Doctoral Thesis. Department of Geography and Regional Science, University of Graz.

Kellerer-Pirklbauer, A. 2013: Ground surface temperature and permafrost evolution in the Hohe Tauern National Park, Austria, between 2006 and 2012: Signals of a warming climate? In: *5th Symposium for Research in Protected Areas – Conference Volume, 10–12 June 2013, Mittersill, Austria*: 363–372.

Kellerer-Pirklbauer, A. & V. Kaufmann 2012: About the relationship between rock glacier velocity and climate parameters in central Austria. *Austrian Journal of Earth Sciences* 105, 2: 94–112.

Kellerer-Pirklbauer, A. & V. Kaufmann (in prep.): Cryospherical and morphodynamical changes in deglaciating cirques in central Austria and its possible significance for permafrost and rock glacier evolution. *Geomorphology*.

Kellerer-Pirklbauer, A., A. Bauer & H. Proske 2005: Terrestrial laser scanning for glacier monitoring: Glaciation changes of the Gößnitzkees glacier (Schober group, Austria) between 2000 and 2004. In: *Proceedings of the 3rd Symposium of the Hohe Tauern National Park for Research in Protected Areas – Conference Volume, 15–17 September 2005, Kaprun, Austria*: 97–106.

Kellerer-Pirklbauer, A., M. Rieckh & M. Avian 2010: Snow-cover dynamics monitored by automatic digital photography at the rooting zone of an active rock glacier in the Hinteres Lantal Cirque, Austria. *Geophysical Research Abstracts* 12: EGU2010-13079.

Kellerer-Pirklbauer, A., G.K. Lieb & H. Kleinferchner 2012: A new rock glacier inventory in the eastern European Alps. *Austrian Journal of Earth Sciences* 105, 2: 78–93.

Kern, K., G.K. Lieb, G. Seier & A. Kellerer-Pirklbauer 2012: Modelling geomorphological hazards to assess the vulnerability of alpine infrastructure: The example of the Großglockner-Pasterze area, Austria. *Austrian Journal of Earth Sciences* 105, 2: 113–127.

Kienast, G. & V. Kaufmann 2004: Geodetic measurements on glaciers and rock glaciers in the Hohe Tauern National Park (Austria). In: *Proceedings of the 4th ICA Mountain Cartography Workshop. Vall de Núria, Catalonia, Spain, 30 September–2 October 2004*: 77–88.

Kneisel, C. & C. Hauck 2008: Electrical methods. In: Hauck, C. & C. Kneisel (eds.): *Applied geophysics in periglacial environments.* Cambridge: 1–27.

Koefoed, O. 1979: *Geosounding Principles, 1: Resistivity sounding measurements.* Methods in Geochemistry and Geophysics 14A. New York.

Krainer, K., A. Kellerer-Pirklbauer, V. Kaufmann, G.K. Lieb, L. Schrott & H. Hausmann 2012: Permafrost Research in Austria: History and recent advances. *Austrian Journal of Earth Sciences* 105, 2: 2–11.

Krysiecki, J.-M., X. Bodin & P. Schoeneich 2008: Collapse of the Bérard Rock Glacier (Southern French Alps). In: Kane, D.L. & K.M. Hinkel (eds.): *Extended Abstracts of the Ninth International Conference on Permafrost*, Fairbanks, Alaska, 29 June–3 July 2008: 153–154.

Lieb, G.K., A. Kellerer-Pirklbauer & H. Kleinferchner 2010: *Rock glacier inventory of Central and Eastern Austria elaborated within the PermaNET project.* Digital Media (Inventory Version 2: January 2012). Institute of Geography and Regional Science, University of Graz.

Rieckh, M. 2011: *Monitoring of the alpine snow cover using automatic digital photography – Results from the Hohe Tauern range (Central Austrian Alps).* Master thesis. Department of Geography and Regional Science, University of Graz.

Rieckh, M., A. Kellerer-Pirklbauer & M. Avian 2011: Evaluation of spatial variability of snow cover duration in a small alpine catchment using automatic photography and terrain-based modelling. *Geophysical Research Abstracts* 13: EGU2011-12048.

Permafrost and Climate Change in North and South Tyrol

Karl Krainer

Institute of Geology, University of Innsbruck

1 Introduction

Permafrost is defined on the basis of temperature, as ground that remains continuously below 0 °C, for at least two consecutive years. Permafrost forms when the ground cools sufficiently in winter to produce a frozen layer that persists throughout the following summer. In the northern hemisphere permafrost is present beneath the surface of 22% of the land area. Permafrost is not restricted to high latitudes but also exists in mountainous mid-latitude areas such as the Alps, called "alpine permafrost".

Permafrost occurs where the mean annual air temperature is near or below 0 °C, typically below a surface zone of annual freeze and thaw called the active layer, mostly 1 to 2 m thick. Frost action changes the composition and structure of the ground, particularly the active layer, by altering soil particles, sorting the particles according to grain size and modifying the shape and structure of the ground surface causing various geomorphic features (summaries in Davies 2001; Yershov 1998; French 1996; Washburn 1979). In the Austrian Alps, alpine permafrost occurs above an altitude of 2,300 to 2,500 m, locally also below. In the Swiss Alps, the area affected by alpine permafrost is estimated to have about the extension of the glacierized area, and the situation may be similar in the Austrian Alps.

Three types of alpine permafrost can be distinguished: a) active rock glaciers, b) permafrost in loose sediments (mainly talus), and c) alpine permafrost in bedrock (fissure ice).

1.1 Rock glaciers

Rock glaciers are debris-covered, slowly creeping mixtures of rock and ice. They transport large amounts of debris downslope with velocities of up to more than 2 m / a. Rock glaciers are common in many alpine and arctic regions (for summary see Barsch 1996; Haeberli 1985, 2005; Whalley & Martin 1992; Haeberli et al. 2006) and belong to the most spectacular and most widespread periglacial phenomenon on earth (Haeberli 1990). Hypotheses about the genesis of rock glaciers have been the subject to long debates and were highly controversial (see discussion for example by Barsch 1996; Haeberli 1985, 1995; Ackert 1998; Clark et al. 1998; Humlum 1996; Johnson 1981; Potter 1972; Potter et al. 1998; Vitek & Giardino 1987; Wahrhaftig & Cox 1959; Whalley & Martin 1992; Whalley et al. 1994; Whalley & Palmer 1998; White 1976). Shroder et al. (2000) proposed that rock glaciers can be formed from glaciers, when sediment is transferred inefficiently from glacier ice to meltwater, based on studies on debris covered glaciers in the Nanga Parbat Himalaya. Recently, Etzelmüller & Hagen (2005) and Haeberli (2005) discussed glacier-permafrost interactions and their relationship in Arctic and high Alpine mountain areas.

Rock glaciers are the most common and most spectacular feature of alpine permafrost (Boeckli et al. 2012). Many rock glaciers in the Alps are located near the lower limit of discontinuous permafrost with temperatures between −2 and 0 °C (Gärtner-Roer et al. 2010).

In the Austrian Alps, many rock glaciers are present (Kellerer-Pirklbauer et al. 2012; Lieb 1996; Lieb et al. 2010), particularly in the Ötztal and Stubai Alps (Gerhold 1967, 1969; Krainer & Ribis 2012). Many of them are exceptionally large and highly active. Most of the total alpine permafrost ice volume is stored in active rock glaciers, while loose sediments and bedrock are considered to contain only minor fractions of the permafrost ice. Detailed investigations on active rock glaciers in the Austrian Alps, their origin and dynamics, have been carried out by the Innsbruck and Graz working groups (e. g. Berger et al. 2004; Brückl et al. 2005; Chesi et al. 1999, 2003; Hausmann et al. 2006a, b, 2007, 2012; Krainer & Mostler 2000a, b, 2001, 2002, 2004, 2006; Krainer et al. 2002, 2007; Lieb 1986, 1987, 1991; Lieb & Slupetzky 1993; Kaufmann 1996a, b, 2012; Kaufmann & Ladstädter 2002, 2003; Kellerer-Pirklbauer & Kaufmann 2012; Ladstädter & Kaufmann 2005; Kienast & Kaufmann 2004; Schmöller & Fruhwirth 1996). A summary of permafrost research in Austria is presented by Krainer et al. (2012).

1.2   Climate change and permafrost

Global average air temperature has increased by more than 0.7 °C between 1906 and 2005, and the decadal warming has almost doubled over the past 50 years with an average value of 0.13 °C per decade (Solomon et al. 2007). The Alpine region has warmed twice as much as the global or Northern Hemispheric mean since the late 19th century, and both mountains and low elevation regions have revealed the same amount of warming (Auer et al. 2007). Global climate models project a temperature increase ranging from 1.8 °C for the low SRES (Special Report Emissions Scenarios) B1 to 4.0 °C for the high scenario A1F1 (Solomon et al. 2007) until the end of this century. Even for the case of a constant radiative forcing, if greenhouse gases and aerosols were kept constant at year 2000 levels, models give a temperature increase of 0.6 °C.

Observations made in Switzerland indicate that the warming during the last 100 years has caused an increase of the lower permafrost boundary by approximately 150 to 250 m, and it is assumed that an increase of the mean annual air temperature of 1 to 2 °C until the middle of the 21st century would cause the equilibrium line of the glaciers to rise by 150 to 350 m, while the lower boundary of the alpine permafrost is expected to rise by 200 to 750 m (Bader & Kunz 1998; Haeberli et al. 1999). In

Switzerland, a monitoring program started in 2000 (PERMOS: Permafrost Monito-ring of Switzerland; see annual reports of PERMOS) to study permafrost temperatu-res and their climate change related variability.

There is still little knowledge on the impact of climate change on melt processes of permafrost ice, on discharge patterns in high alpine regions, and on water quality. First data on the hydrologic regime and discharge of active rock glaciers were publis-hed by Krainer & Mostler (2002), and Krainer et al. (2007).

In areas of ice-rich permafrost, discharge during the melting season is expected to increase as response to climate warming. Enhanced melting of permafrost ice may raise the suspended load in melt water released from active rock glaciers thus incre-asing the input of fine-grained sediment into reservoir basins. Extremely high con-centrations of Nickel (*Ni*), which strongly exceeded the limit of drinking water, were determined in meltwater released from active rock glaciers and glaciers at Kaunertal and Schnalstal (Ötztal Alps) (unpublished data). Increasing concentrations of ions and heavy metals were found in two high Alpine lakes, which are impacted by melt water from rock glaciers (Thies et al. 2007). Climate change induced permafrost de-gradation may have major impacts on ecosystems, landscape stability and on people and their livelihoods.

In particular, high concentrations of *Ni* may strongly exceed the limit of drinking water as has been detected in melt water derived from active rock glaciers and glaciers at Schnalstal (Ötztal Alps in South Tyrol, Italy) (Mair et al. 2011). Only recently studies have been published focussing on the impact of alpine permafrost in uncon-solidated sediments on the hydrological regime (Clow et al. 2003; Liu et al. 2004; McClymont et al. 2011; Rogger et al. in prep).

1.3   The aims of the project are

- to study and document the ice content of alpine permafrost (particularly of ac-tive rock glaciers)
- to assess the impact of climate change (i. e. global warming) on
  - › melting of permafrost ice (particularly active rock glaciers)
  - › discharge patterns in high Alpine regions
  - › water chemistry of melt water released from alpine permafrost (particularly from active rock glaciers) into surface and groundwater (e. g. drinking water supplies)
- to evaluate the regional distribution in the occurrence of increasing ion concen-trations and elevated heavy metal values (e. g. *Ni*, *Mn*) in drainage waters from active rock glaciers across the Tyrolean Alps
- to evaluate potential sources of elevated solute and heavy metal values
- to provide information on drinking water supplies impacted by high nickel values as derived from melting rock glaciers
- to provide basic data for biological studies (impact of high *Ni*, *Mn* concentrations on ecology, biological processes)

1.4   Distribution of permafrost and ice content in the western Austrian Alps

Permafrost is widespread in the European Alps and includes a large number of rock glaciers, which are the typical and most common permafrost landform and particularly abundant in the Tyrolean Alps (Austria).

A data collection sheet for a rock glacier inventory of Tyrol and Vorarlberg was created which contains the following data: number (according to the catchment), geographical name, coordinates, elevation of the front, rooting zone and average height, maximum length and width, area, aspect, surface morphology, shape, origin, status (active, inactive, fossil), water-catchment, mountain range, bedrock in the catchment area, springs at the base of the front, information on existing water analysis, discharge data, literature. The determination of these data from aerial images is often limited or even impossible. The distinction between active, inactive and fossil made from aerial photographs is difficult, as there are smooth transitions between these types. The current state of a rock glacier usually can be detected only by flow velocity measurements and other tests.

The data provide an important basis for estimating the distribution of permafrost and related hydrological processes (enhanced melting of permafrost ice and its impacts on the runoff) and natural hazards (debris flows). We compiled a rock glacier inventory of all mountain groups of western Austria (Tyrol, Vorarlberg). Each rock glacier is documented by an orthophoto and by a datasheet which contains information such as coordinates, altitude, length, width, area, exposition (flow direction), shape, state, hydrology and bedrock in the catchment area. All rock glaciers are listed in an excel-sheet. The inventory is based on the study of high-quality aerial photographs and laser scan images. The rock glacier inventory of the Tyrolean Alps includes 3,145 rock glaciers which cover an area of 167.2 km² (Fig. 1). Of these, 517 (16.4%) were classified as active, 915 (29.1%) as inactive, and 1,713 (54.5%) as fossil (Krainer & Ribis 2012).

Tongue-shaped, talus-derived, ice-cemented rock glaciers are the most common type among active and inactive rock glaciers. Glacier-derived rock glaciers containing a massive ice-core are rare. Most rock glaciers occur in the mountain groups of the central Alps in which bedrock is composed mainly of mica schists, paragneiss, orthogneiss and amphibolites ("Altkristallin"). The majority of active and inactive rock glaciers are exposed towards a northern (NW, N and NE) direction. Active and inactive rock glaciers exposed towards S, SE and SW are minimal.

The average ice content of the frozen core drilled at rock glacier Lazaun was 22% at core Lazaun I (near the front), 43% at core Lazaun II, and probably higher in the rooting zone of the rock glacier (Krainer et al. submitted). From geophysical data Hausmann et al. (2007, 2012) calculated ice contents of 45 to 60% for the lower part of Reichenkar rock glacier, 43 to 61% for Ölgrube rock glacier and 45 to 60% for Kaiserberg rock glacier. The total amount of ice in active and inactive rock glaciers is estimated to be 0.19 to 0.27 km³ which is small compared to the ice volume contained in the glaciers of the Tyrolean Alps.

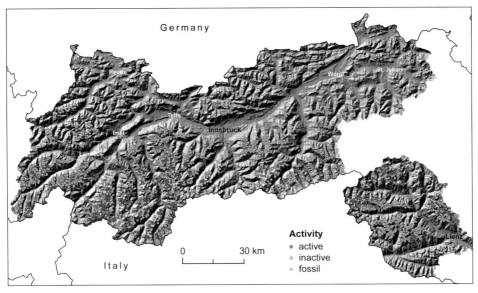

Figure 1: Distribution of rock glaciers in the Tyrolean Alps (blue circles: active r.g., orange: inactive r.g., green: fossil r.g.). Rock glaciers are most abundant in the mountain groups of the central Alps which are composed of metamorphic rocks such as schists and gneisses (Silvretta and Samnaun Mountain Groups, Ötztal and Stubai Alps, Deferegger Alps and Schober Mountain Group).

The distribution of active and inactive rock glaciers indicates that the lower limit of discontinuous permafrost in the mountain groups located in the central part of the Tyrolean Alps of Austria is located at approximately 2,500 m (Krainer & Ribis 2012).

In the westernmost part of the Austrian Alps (Vorarlberg) 202 rock glaciers were recorded which cover an area of approximately 7 km². Most of the rock glaciers are located in the Verwall Mountain Group (97) and Silvretta Mountain Group (57), both mainly composed of metamorphic rocks. Most of the rock glaciers in Vorarlberg (145 or 71.8%) are classified as fossil, covering an area of approximately 6 km². Inactive rock glaciers are less common (46 or 22.8%), they cover an area of 0,93 km². Only 11 rock glaciers (5.4%) are classified to be slightly active covering an area of 0,19 km² (Stocker 2012).

The distribution of alpine permafrost in unconsolidated sediment was studied in detail in a small catchment in the western part of the Ötztal Alps (details in Krainer & Hausmann 2013; Hausmann et al. in prep). The study site was Krummgampen Valley, a small side valley of Kaunertal. The area of the catchment measures 5.76 km², the elevation in the catchment area extends from 2 400 m to the highest summit (Glockturm) at 3,350 m. Investigations in the catchment area include detailed mapping, grain-size analysis of selected sediment samples, ground surface temperatures, hydrological investigations (discharge, water temperature, electrical conductivity) and geophysical surveying (refraction seismics, ground penetrating radar).

Data show that approximately 71% of the catchment area is underlain by permafrost (20 % discontinuous and 51% sporadic permafrost). Discontinuous permafrost occurs in bedrock (7%), till deposits of the Little Ice Age (LIA) (5%), rock glaciers (4%) and talus slope (4%). Sporadic permafrost is most common in bedrock (19%), LIA till deposits (18%), talus slope (11%) and rock glaciers (0.6%). Discontinuous permafrost is abundant in areas (slopes) facing towards a northern direction. Permafrost also occurs locally in talus slopes facing towards south at elevations of approximately 2,750 m which is indicated by geophysical data and ground surface temperatures with the lowest temperatures (–6 °C) on coarse-grained sediment at the front of talus slopes. No permafrost was determined on pre-LIA till deposits.

2   Information from core drilling on an active rock glacier

Important data were obtained from two cores which were drilled at rock glacier Lazaun located at Lazaunkar west of Kurzras in the southern Ötztal Alps (Schnals Valley, South Tyrol). A detailed description and discussion of this rock glacier including the core drillings is presented by Krainer et al. (submitted).

Rock glacier Lazaun is a medium-sized active, tongue-shaped rock glacier with a steep front with gradients of 30 to 50°. The rock glacier is 660 m long and up to 200 m wide (Fig. 2). The depression in the rooting zone indicates melting of a massive ice core in this part of the rock glacier. Flow velocity measurements, bottom temperature of snow cover (BTS), water temperature of the springs, steep front and surface morphology demonstrate that the rock glacier is active and contains substantial amounts of permafrost ice.

The discharge pattern is typical for active rock glaciers and is characterized by strong diurnal and seasonal variations. During winter (October until May) discharge is extremely low and electrical conductivity high. Highest discharge is recorded during the snowmelt period in June and July, and during rainfall events. Pronounced diurnal variations in discharge are recorded in May and June. From the end of July until October discharge decreases, interrupted by single peaks caused by rainfall events. Warm weather periods in autumn may also cause a slight increase in discharge (increased melting of permafrost ice).

The rock glacier spring is characterized by relatively high electrical conductivity with values of 100 to 275 µS / cm. Water temperature of the rock glacier spring is low (1.3 °C or less) during the entire melt season.

Data from the two cores indicate that the average ice content of the rock glacier is approximately 35 to 40 vol.%. The frozen core of the rock glacier covers an area of approximately 0.1 km², the annual melting rate of the rock glacier ice according to GPS measurements is in the order of 10 cm on average resulting in a total ice volume of 10,000 m³ (approx. 9,100 m³ water) which the rock glacier loose by melting each year during the melt season from May until October (six months). This results in an average discharge of 0.6 l / s which is only about 2.3% of the average discharge of the rock glacier (approx. 26 l / s).

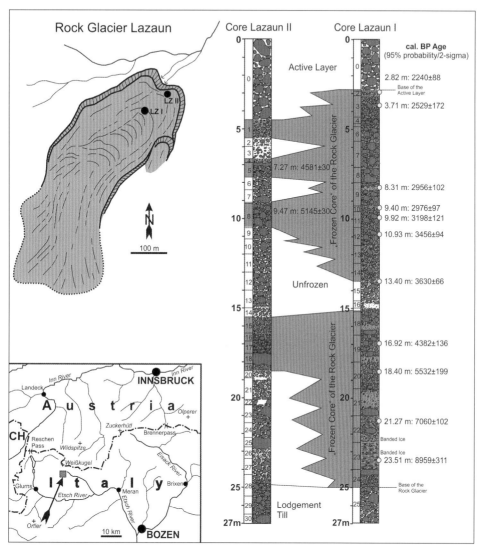

Figure 2: Location map and map showing the morphology and location of the core drillings at active rock glacier Lazaun. On the right are graphic logs through the cores Lazaun I and Lazaun II showing frozen parts (gray) and C¹⁴-ages

This indicates that the amount of melt water derived from the melting of permafrost ice is very low. Even if the melting rate of permafrost ice is 20 cm / year, the amount is less than 5% of the total discharge of the rock glacier. Discharge is mostly derived from snowmelt and summer rainfall with very small amounts of groundwater and melting of permafrost ice.

Core Lazaun I is 40 m long. The active layer (unfrozen debris layer) is 2.8 m thick and underlain by a continuous frozen core (mixture of ice and debris) from 2.8 m to a depth of 13.5 m. From 13.5 to 15.2 m almost ice-free debris is present, and from 15.2 to 25 m again a continuous frozen core was obtained. From 19.5 to 25 m the core contained high amounts of dark colored, banded ice and fine-grained sediment with few clasts with diameters up to several cm floating in the banded ice (Fig. 2 and 3). From 25 to 28 m coarse debris is present, underlain by debris with high amounts of fine-grained sediment down to a depth of 40 m.

Core Lazaun II is 32 m long. This core which was drilled close to the front of the rock glacier contains significantly lower amounts of ice compared to core Lazaun I (Figure 2). The active layer is 4.5 m thick. Ice is present from 4.5 to 5.5 m, from 6.65 to 7.7 m, from 9.1 to 10.5 m and from 15.5 to 18.5 m. From 18.5 to 24.5 m coarse debris with small amounts of fine sediment was obtained, whereas from 24.5 to 32 m coarse debris with high amounts of fine sediment occurred. The average ice content of the frozen core is 22%.

The ice content of the frozen core varies considerably from almost zero up to 98%. The average ice content of core Lazaun I between 2.8 and 25 m is 43%. The average ice content is higher between 2.8 and 14 m (48%) and between 19.5 and 25 m (51%). We assume that in the upper part of the rock glacier, particularly in the rooting zone the ice content is somewhat higher. Near the front (core Lazaun II) the ice content is 22%.

The average ice content of Lazaun rock glacier is approximately 35 to 40%. The total area of the rock glacier is 0.12 km², the area of the frozen core is estimated to be 0.1 km² resulting in a total ice volume of 740,000 to 850,000 m³ (average thickness of the frozen core 21 m).

The melting rate of ice of Lazaun rock glacier is in the order of 10 cm / a resulting in a total loss of ice by melting of 10,000 m³ / year. Thus at present melting of permafrost ice causes a loss of about 1.2 to 1.3% of the total ice volume each year.

At the base of core Lazaun I banded ice with low amounts of debris is present. The rock glacier is composed of two frozen bodies, separated by an unfrozen debris layer which is about 2 m thick. At present both frozen bodies are active. Inclinometer measurements indicate that deformation occurs within a shear horizon at a depth of 20 to 24 m which is at the base of the lower frozen core, and to a minor extent also at a shear horizon at 14 m which is at the base of the upper frozen core.

Borehole temperatures in the frozen core are near the melting point never decreasing below –0.9 °C, indicating the presence of warm permafrost ice. Although no ice was determined below a depth of 24 m, borehole temperatures in the debris layer and lodgement till are slightly below zero down to a depth of 35 m.

Radiocarbon ages from core Lazaun I range from 2,235 cal. BP at a depth of 2.82 m (near the surface of the frozen core of the rock glacier) to 8,959 ± 311 cal. BP at a depth of 23.51 m, approximately 1.5 m above the base of the frozen core of the rock glacier (Fig. 2). One sample from core Lazaun II from a depth of 9.5 m yielded an age of 5,145 ± 30 cal. BP. Radiocarbon ages indicate that the ice at the base of the rock glacier is approximately 10,000 years old and that the frozen core of the rock glacier represents an undisturbed stratigraphic succession covering a time span from

Figure 3: Frozen core from the drill hole Lazaun I at a depth of 18 to 18.5 m composed of banded ice and debris

10,000 to 2,200 years BP. Radiocarbon ages demonstrate that the rock glacier started to form about 10 000 years ago and since this time was intact. Therefore, periods of activity must have alternated with periods of inactivity. Even during warm periods the ice of the rock glacier persisted although the temperature of the permafrost ice at present is close to the melting point. The rock glacier underlain by lodgement till which can be ascribed to the Egesen (Younger Dryas).

3   Permafrost, water chemistry and heavy metal concentrations in melt water

High *Ni* concentrations were first recorded at a high Alpine lake (Rassass See) in South Tyrol (Thies et al. 2007). In the meantime we recognized a number of springs derived from active rock glaciers which contain high concentrations of *Ni* and other heavy metals such as *Co, Cu, Fe, Mn* and *Zn* (Krainer et al. 2012; Nickus et al. in prep). In the cirque of Lazaunalm (Schnals Valley, Ötztal Alps, South Tyrol) all springs derived from active rock glaciers and glaciers are characterized by low temperature, high electric conductivity (100 to 275 µS / cm) and high concentrations of *Ni, Fe, Mn* and *Zn*. Electric conductivity and heavy metal concentrations are lowest during the peak snowmelt period and increase towards autumn (up to 0.175 mg/l *Ni*) indicating that the heavy metals are derived from melting of permafrost ice and glacier ice.

Analysis of the ice of the frozen core Lazaun I showed that heavy metals, particularly *Ni* are enriched in several levels in the upper part of the frozen core with peaks up to 0.49 mg/l *Ni*, 0.705 mg/l *Zn*, 0.43 mg/l *Co*, 0.095 mg/l *Cu*, 16.226 mg/l *Fe* and 4.826 mg/l *Mn*. Less high are the concentrations in the lower part of the core between 15 and 24 m showing maximum values of 0.054 mg/l *Ni*, 0.030 mg/l *Zn*, 0.060 mg/l *Co*, 0.095 mg/l *Cu*, 1.383 mg/l *Mn* (Fig. 4).

High heavy metal concentrations were also determined in the ice at distinct levels of core Lazaun II. Analyses of the bedrock indicate that *Ni* and other heavy metals are not derived from the rocks in the catchment area of the rock glacier which are

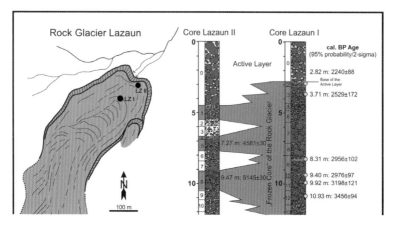

Figure 4: The distribution of Ni in the frozen core Lazaun I. High Ni concentrations were recorded in the upper part with peaks showing extreme values at depths of 4.1 m, 9.8 m and 12.2 m

composed of mica schist and paragneiss of the Ötztal-Stubai-Metamorphic Complex. Analyses of the permafrost ice also demonstrate that the high *Ni* concentrations in the spring water are derived from distinct horizons of the permafrost ice.

High *Ni* concentrations were measured in a 2 m thick ice core which was drilled on active rock glacier Rossbänk (Ulten Valley, South Tyrol). High *Ni* concentrations are also recorded in rock glacier springs at Inneres Hochebenkar near Obergurgl, in the Windach Valley south of Sölden, at Krummgampental and Wurmetal (Kaunertal), all located in the Ötztal Alps (Tyrol, Austria) and also in a small creek derived from a glacier south of the Franz Senn Hütte (Stubai Alps) (Fig. 5). The source of the *Ni* is unknown and we assume that *Ni* is derived from the atmosphere.

Water released from Hochebenkar rock glacier has been studied for its chemical composition since summer 2007 (Nickus et al. in press). At one of the rock glacier springs, which is located on the eastern side at an elevation of 2,560 m the electrical conductivity is high. The values range between 100 and 200 µS / cm during the major snow melt period in June and increase with decreasing runoff. In fall, solute concentration reaches its maximum, and electrical conductivity is around 500 µS / cm. Heavy summer precipitation events generally cause a dilution of the highly concentrated water of the spring, and runoff peaks often coincide with low electrical conductivity values. Sulfate, calcium and magnesium dominate the ion content and comprise more than 90% of the ion balance. Heavy metals are absent.

The seasonal variation of the solute concentration reflects the varying contributions of snowmelt, precipitation events, groundwater and melting of permafrost ice to the runoff. The authors assume that the high amount of solutes in late summer and fall is released from the permafrost ice of the rock glacier. Melt water from permafrost ice seems to be particularly rich in sulfate and the relative contribution of sulfate to the total ion content of the rock glacier runoff generally rises from late spring to fall (Nickus et al. in prep.).

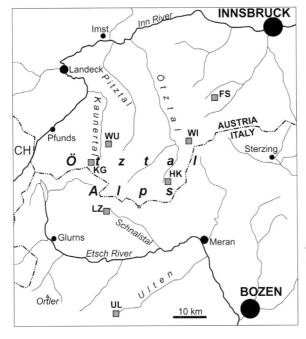

Figure 5: Map showing the location of rock glacier springs and glacier creeks containing high concentrations of Ni (UL Ulten Valley, Ni in the ice of the frozen core; LZ Lazaun, Ni in rock glacier springs and glacier creek; KG Krummgampen, Ni in rock glacier springs; WU Wurmetal, Ni in springs derived from permafrost; WI Windach, Ni in several rock glacier springs; FS Franz Senn Hütte, Ni in a glacier creek)

4   Permafrost and discharge

The influence of climate change on the discharge of catchment areas underlain by permafrost is one of the main goals. Hydrological studies on active rock glaciers showed that discharge is mainly controlled by the local weather conditions, the thermal properties of the unfrozen debris layer (active layer), and the physical mechanisms that control the flow of melt water through the rock glacier (Krainer & Mostler 2002; Krainer et al. 2007).

In general, discharge of active rock glaciers is characterized by strong seasonal and diurnal variations (Fig. 6). Most of the water released at rock glacier springs is derived from snowmelt and rainfall events, and only small amounts are derived from melting of permafrost ice and groundwater. Water derived from snowmelt and rainfall events during summer is quickly released producing sharp discharge peaks. Fair weather periods with intense melting of snow cause pronounced diurnal variations in discharge during the snowmelt period (May to July). Water temperature of springs of active rock glaciers remains constantly around 1 °C during the entire melt period. Values of delta[18]O and electrical conductivity (EC) of the melt water are lowest during high discharge, particular during the main snowmelt period. Delta[18]O and EC progressively increase until late July to early August when the snow of the preceding winter is completely melted. Highest values of EC are recorded in autumn. This increase in delta[18]O and EC is caused by a progressive decrease in the ratio of snowmelt versus ice melt plus groundwater. Meltwater derived from summer rain-

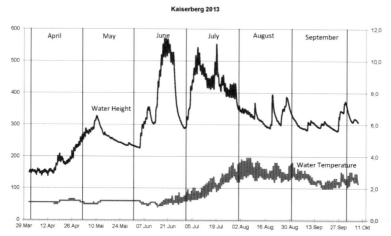

Figure 6: Discharge pattern of an active rock glacier (Kaiserberg, western Ötztal Alps) for the period April to October 2013. The melt season started in mid-April, discharge was relatively low during May and the beginning of June due to cool weather, increased strongly in mid-June caused by extremely warm weather, decreased at the end of June due to cool weather, was high at the beginning of July and then continuously decreased towards October just interrupted by single peaks caused by rainfall events (left scale: water height in mm, right scale: water temperature in °C)

fall events is quickly released within a few hours causing sharp peaks in discharge, a less pronounced peak in delta[18]O and a pronounced decrease in electric conductivity (Krainer & Mostler 2002; Krainer et al. 2007).

Gobal Positioning System (GPS) measurements and hydrological data indicate the annual melting rate of permafrost ice of rock glaciers is small and that the amount of water released by the melting of permafrost ice is less than 5% of the total discharge.

To understand the subsurface flow paths in a permafrost-influenced catchment and to assess the effect of increased melting of permafrost on the hydrologic regime, detailed hydrological studies were carried out at Krummgampen the distribution of permafrost was studied in detail (see above). A distributed hydrological model was applied to simulate the discharge in the catchment of Krummgampen under the present conditions and in a future scenario without permafrost. The simulations indicate that the complete melting of permafrost ice in the catchment of Krummgampen will increase the storage capacity of the sediments which will reduce the flood peaks up to 20% and increase runoff during recession of about 15% (details in Krainer & Hausmann 2013; Rogger et al. in prep.).

5   Conclusions

The high number of active and inactive rock glaciers in the western part of the Austrian Alps documents that alpine permafrost is widespread, particularly in mountain groups in the central part of the Austrian Alps which are composed of metamorphic rocks (schists and gneisses). Detailed permafrost mapping at Krummgampen shows that permafrost is not restricted to active and inactive rock glaciers but is also common in unconsolidated sediments such as talus and till deposits, particularly on north-facing slopes above an elevation of 2,500 m.

Radiocarbon dating of ice from active rock glacier Lazaun shows that permafrost ice of rock glaciers may be up to 10,000 years old. Although borehole measurements at rock glacier Lazaun demonstrate that the temperature of the ice is close to the melting point, the ice persisted even warmer periods during the last 10,000 years.

Hydrological studies show that the amount of melt water released from rock glaciers due to increased melting caused by global warming is small compared to the total discharge. Most of the melt water is derived from snowmelt and precipitation and less than 5% is derived from melting of permafrost ice.

Chemical analyses of melt water show that some rock glacier springs are highly contaminated by *Ni* and other heavy metals. Analyses of the ice from the core drilled at rock glacier Lazaun demonstrate that *Ni* is concentrated in the permafrost ice and that individual horizons in the permafrost ice contain extremely high amounts of *Ni* and other heavy metals. We assume that not only at rock glacier Lazaun but also at other locations *Ni* and other heavy metals in the melt water are derived from increased melting of permafrost ice. The source of *Ni* is unknown, chemical analyses indicate that *Ni* is not derived from the bedrock.

Increased melting of permafrost ice in rock glaciers and unconsolidated sediments such as talus and till deposits will reduce the volume of permafrost ice and increase the pore space and thus the storage capacity for water in the sediments. Hydrological simulations at Krummgampen indicate that increased melting of permafrost will result in a decrease in flood peak discharge of up to 20% and an increase of runoff during recession periods of up to 15% due to an increase in the storage capacity of the sediments in the catchment.

6   Acknowledgements

The present paper is a brief summary of results obtained from scientific projects funded by the Austrian Academy of Sciences (International Strategy for Desaster Reduction, ISDR), FWF (Austrian Science Fund) and PermaNET. The author wishes to thank all persons involved in these projects, particularly G. Blöschl, E. Brückl, H. Hausmann, U. Nickus, M. Rogger, R. Tessadri and H. Thies.

7   References

Ackert, R.P. 1998: A rock glacier / debris-covered glacier system at Galena Creek, Absaroka Mountains, Wyoming. *Geografiska Annaler* 80, 3-4: 267–276.

Auer, I., R. Böhm, A. Jurkovic, W. Lipa, A. Orlik, R. Potzmann, W. Schöner, M. Ungersböck, C. Matulla, K. Briffa, P. Jones, D. Efthymiadis, M. Brunetti, T. Nanni, M. Maugeri, L. Mercalli, O. Mestre, J.-M. Moisselin, M. Begert, G. Müller-Westermeier, V. Kveton, O. Bochnicek, P. Stastny, M. Lapin, S. Szalai, T. Szentimrey, T. Cegnar, M. Dolinar, M. Gajic-Capka, K. Zaninovic, Z. Majstorovic & E. Nieplova 2007: HISTALP-historical instrumental climatological surface time series of the Greater Alpine Region. *International Journal of Climatology* 27: 17–46.

Barsch, D. 1996: *Rockglaciers. Indicators for the Present and Former Geoecology in High Mountain Environments*. Berlin.

Berger, J., K. Krainer & W. Mostler 2004: Dynamics of an active rock glacier (Ötztal Alps, Austria). *Quaternary Research* 62: 233–242.

Boeckli, L., A. Brenning, S. Gruber & J. Noetzli 2012: A statistical approach to modelling permafrost distribution in the European Alps or similar mountain ranges. *The Cryosphere* 6: 125–140. doi:10.5194/tc-6-125-2012.

Brückl, E., H. Hausmann, K. Krainer & W. Mostler 2005: Internal structure of Reichenkar rock glacier. *Geophysical Research Abstracts* 7, SRef-ID: 1607-7962/gra/EGU05-A-02358.

Chesi, G., K. Krainer, W. Mostler & T. Weinold 1999: Bewegungsmessungen am aktiven Blockgletscher Inneres Reichenkar mit der GPS-Methode. In: *10. Internationale Geodätische Woche Obergurgl 1999*: 223–227.

Chesi, G., S. Geissler, K. Krainer, W. Mostler & T. Weinold 2003: 5 Jahre Bewegungsmessungen am aktiven Blockgletscher Inneres Reichenkar (westliche Stubaier Alpen) mit der GPS-Methode. In: *12. Internationale Geodätische Woche Obergurgl 2003*: 201–205.

Clark, D.H., E.J. Steig, N. Potter & A.R. Gillespie 1998: Genetic variability of rock glaciers. *Geografiska Annaler* 80, 3-4: 175–182.

Clow, D.W., L. Schrott, R. Webb, D.H. Campbell, A. Torizzo & M. Domblaser 2003: Ground Water Occurrence and Contributions to Streamflow in an Alpine Catchment, Colorado Front Range. *Ground Water – Watershed* 41, 7: 937–950. doi:10.1111/j.1745-6584.2003.tb02436.x.

Davis, N. 2001: *Permafrost: a guide to frozen ground in transition*. Fairbanks.

Etzelmüller, B. & J.O. Hagen 2005: Glacier – permafrost interaction in Arctic and alpine mountain environments with examples from southern Norway and Svalbard. In: Harris, C. & J.B. Murton (eds.): *Cryospheric Systems: Glaciers and Permafrost*. Geological Society Special Publication 242. London: 11–27.

French, H.M. 1996: *The Periglacial Ennvironment*. Essex.

Gärtner-Roer, I. 2010: Permafrost. In: Voigt, T., H.-M. Füssel, I. Gärtner-Roer, C. Huggel, C. Marty & M. Zemp (eds.): *Impacts of climate change on snow, ice, and permafrost in Europe: Observed trends, future projections, and socioeconomic relevance*. ETC/ACC Technical Paper 2010/13. Bilthoven: 66–76.

Gerhold, N. 1967: Zur Glazialgeologie der westlichen Ötztaler Alpen. *Veröffentlichungen des Museum Ferdinandeum* 47: 5–50.

Gerhold, N. 1969: Zur Glazialgeologie der westlichen Ötztaler Alpen unter besonderer Berücksichtigung des Blockgletscherproblems. *Veröffentlichungen des Museum Ferdinandeum* 49, 45–78.

Haeberli, W. 1985: Creep of mountain permafrost: Internal structure and flow of alpine rock glaciers. *Mitteilungen der Versuchsanstalt für Wasserbau, Hydrologie und Glaziologie ETH Zürich* 77: 1–142.

Haeberli, W. 1990: Scientific, environmental and climatic significance of rock glaciers. *Memorie della Società Geologica Italiana* 45: 823–831.

Haeberli, W. 1995: Permafrost und Blockgletscher in den Alpen. *Vierteljahrsschrift der Naturforschenden Gesellschaft in Zürich* 140, 3: 113–121.

Haeberli, W. 2005: Investigating glacier – permafrost relationships in high-mountain areas: historical background, selected examples and research needs. In: Harris, C. & J.B. Murton (eds.): *Cryospheric Systems: Glaciers and Permafrost*. Geological Society Special Publication 242. London: 29–37.

Haeberli, W., B. Hallet, L. Arenson, R. Elconin, O. Humlum, A. Kääb, V. Kaufmann, B. Ladanyi, N. Matsuoka, S. Springman & D. Vonder Mühll 2006: Permafrost Creep and Rock Glacier Dynamics. *Permafrost and Periglacial Processes* 17: 189–216.

Hausmann H., K. Krainer & E. Brückl (submitted). Mapping and modelling of mountain permafrost using seismic refraction and ground surface temperatures, Krummgampen Valley, Ötztal Alps, Austria.

Hausmann, H., K. Krainer, E. Brückl & W. Mostler 2006a: Dynamics of Alpine rock glaciers in the context of global warming. *Geophysical Research Abstracts* 8, SRef-ID: 1607-7962/gra/EGU06-A-04718.

Hausmann, H., K. Krainer, E. Brückl & W. Mostler 2006b: Creep of three alpine rock glaciers – observation and modelling (Ötztal and Stubai Alps, Austria). In: *Abtracts of the HMRSC-IX. 9th International Symposium on High Mountain Remote Sensing Cartography, Graz, Austria, 14–22 September 2006:* 60.

Hausmann, H., K. Krainer, E. Brückl & W. Mostler 2007: Internal structure, composition and dynamics of Reichenkar rock glacier (western Stubai Alps, Austria). *Permafrost and Periglacial Processes* 18, 351–367. doi:10.1002/ppp.60.

Hausmann, H., K. Krainer, E. Brückl & C. Ullrich 2012: Internal structure, ice content and dynamics of Ölgrube and Kaiserberg rock glaciers (Ötztal Alps, Austria) determined from geophysical surveys. *Austrian Journal of Earth Sciences* 105, 2: 12–31.

Humlum, O. 1996: Origin of Rock Glaciers: Observations from Mellemfjord, Disko Island, Central West Greenland. *Permafrost and Periglacial Processes* 7: 361–380.

Johnson, P.G. 1981: The structure of a talus-derived rock glacier deduced from its hydrology. *Canadian Journal of Earth Sciences* 18: 1422–1430.

Kaufmann, V. 1996a: Der Dösener Blockgletscher – Studienkarten und Bewegungsmessungen. In: *Beiträge zur Permafrostforschung in Österreich, Arbeiten aus dem Institut für Geographie, Karl-Franzens-Universität Graz* 33: 141–162.

Kaufmann, V. 1996b: Geomorphometric monitoring of active rock glaciers in the Austrian Alps. In: *Proceedings of the 4th international symposium on High Mountain Remote Sensing Cartography. Karlstad, Kiruna, Tromso, 19–29 August 1996:* 97–113.

Kaufmann, V. 2012: The evolution of rock glacier monitoring using terrestrial photogrammetry: the example of Äußeres Hochebenkar rock glacier (Austria). *Austrian Journal of Earth Sciences* 105, 2: 63–77.

Kaufmann, V. & R. Ladstädter 2002: Spatio-temporal analysis of the dynamic behavioud of the Hochebenkar rock glaciers (Oetztal Alps, Austria) by means of digital photogrammetric methods. *Grazer Schriften der Geographie und Raumforschung* 37: 119–140.

Kaufmann, V. & R. Ladstädter 2003: Quantitative analysis of rock glacier creep by means of digital photogrammetry using multi-temporal aerial photographs: two case studies in the Austrian Alps. In: Phillips M., S.M. Springman & L.U. Arenson (eds.): *Permafrost: Proceedings of the 8th International Conference on Permafrost*, Zurich, Switzerland, 21-25 July 2003: 525–530.

Kellerer-Pirklbauer, A., G.K. Lieb & H. Kleinferchner 2012: A new rock glacier inventory of the Eastern European Alps. *Austrian Journal of Earth Sciences* 105, 2: 78–93.

Kienast, G. & V. Kaufmann 2004: Geodetic measurements on glaciers and rock glaciers in the Hohe Tauern National Park (Austria). In: *Proceedings of the 4th ICA Mountain Cartography Workshop. Vall de Núria, Catalonia, Spain, 30 September–2 October 2004:* 77–88.

Krainer, K. & H. Hausmann 2013: *Permafrost in Austria: Impact of climate change on alpine permafrost and related hydrological effects. Austrian Academy of Sciences*, ISDR Final Report 2007–2011. doi: 10.1553/ISDR-22s1.

Krainer, K. & W. Mostler 2000a: Reichenkar Rock Glacier: a Glacier Derived Debris-Ice System in the Western Stubai Alps, Austria. *Permafrost and Periglacial Processes* 11: 267–275.

Krainer, K. & W. Mostler 2000b: Aktive Blockgletscher als Transportsysteme für Schuttmassen im Hochgebirge: Der Reichenkar Blockgletscher in den westlichen Stubaier Alpen. *Geoforum Umhausen* 1: 28–43.

Krainer, K. & W. Mostler 2001: Der aktive Blockgletscher im Hinteren Langtal Kar, Gößnitztal (Schobergruppe, Nationalpark Hohe Tauern, Österreich). *Wissenschaftliche Mitteilungen aus dem Nationalpark Hohe Tauern* 6: 139–168.

Krainer, K. & W. Mostler 2002: Hydrology of Active Rock Glaciers: Examples from the Austrian Alps. *Arctic, Antarctic, and Alpine Research* 34: 142–149.

Krainer, K. & W. Mostler 2004: Aufbau und Entstehung des aktiven Blockgletschers im Sulzkar, westliche Stubaier Alpen (Tirol). *Geo.Alp* 1: 37–55.

Krainer, K. & W. Mostler 2006: Flow Velocities of Active Rock Glaciers in the Austrian Alps. *Geografiska Annaler* 88A: 267–280.

Krainer, K. & M. Ribis 2012: A Rock Glacier Inventory of the Tyrolean Alps (Austria). *Austrian Journal of Earth Sciences* 105, 2: 32–47.

Krainer, K., W. Mostler & N. Span 2002: A glacier-derived, ice-cored rock glacier in the western Stubai Alps (Austria): evidence from ice exposures and ground penetrating radar investigation. *Zeitschrift für Gletscherkunde und Glazialgeologie* 38: 21–34.

Krainer, K., W. Mostler & C. Spötl 2007: Discharge from active rock glaciers, Austrian Alps: a stable isotope approach. *Austrian Journal of Earth Sciences* 100: 102–112.

Krainer, K., A. Kellerer-Pirklbauer, V. Kaufmann, G.K. Lieb, L. Schrott & H. Hausmann 2012: Permafrost Research in Austria: History and recent advances. *Austrian Journal of Earth Sciences* 105, 2: 2–11.

Krainer, K., K. Lang, V. Mair, U. Nickus, R. Tessadri, D. Tonodandel & H. Thies 2012: Core drilling on active rock glacier Lazaun (southern Ötztal Alps, South Tyrol). In: *Pangeo Austria, Salzburg, 15–20 September 2012 – Abstracts*: 83–84.

Krainer, K., D. Bressan, B. Dietre, J.N. Haas, I. Hajdas, U. Nickus, D. Reidl, H. Thies & D. Tonidandel (submitted): A 10 300-year old ice core from active rock glacier Lazaun, southern Ötztal Alps (South Tyrol, northern Italy).

Ladstädter, R. & V. Kaufmann 2005: Studying the movement of the Outer Hochebenkar rock glacier: Aerial vs. ground-based photogrammetric methods. In: *2nd European Conference on Permafrost, Potsdam, Germany, 12–16 June 2005, Programme and Abstracts*. Terra Nostra, 2005 / 2: 97.

Lieb, G.K. 1996: Permafrost und Blockgletscher in den östlichen österreichischen Alpen. In: Institut für Geographie der Karl-Franzens-Universität Graz (Hrsg.): *Beiträge zur Permafrostforschung in Österreich*. Grazer Schriften der Geographie und Raumforschung 33: 9–125.

Lieb, G.K. 1986: Permafrost und Blockgletscher der östlichen Schobergruppe (Hohe Tauern, Kärnten). In: Institut für Geographie der Karl-Franzens-Universität Graz (Hrsg.): *Festschrift für Wilhelm Leitner zum 60. Geburtstag*. Grazer Schriften der Geographie und Raumforschung 27: 123–132.

Lieb, G.K. 1987: Zur spätglazialen Gletscher- und Blockgletschergeschichte im Vergleich zwischen den Hohen und Niederen Tauern. *Mitteilungen der Österreichischen Geographischen Gesellschaft* 129: 5–27.

Lieb, G.K. 1991: Die horizontale und vertikale Verteilung der Blockgletscher in den Hohen Tauern (Österreich). *Zeitschrift für Geomorphologie N.F.* 35, 3: 345–365.

Lieb, G.K. & H. Slupetzky 1993: Der Tauernfleck-Blockgletscher im Hollersbachtal (Venedigergruppe, Salzburg, Österreich). *Wissenschaftliche Mitteilungen aus dem Nationalpark Hohe Tauern* 1: 138–146.

Lieb, G.K., A. Kellerer-Pirklbauer & H. Kleinferchner 2010: *Rock glacier inventory of Central and Eastern Austria elaborated within the PermaNET project*. Digital Media (Inventory Version 2: January 2012). Institute of Geography and Regional Science, University of Graz.

Liu, F., M.W. Williams & N. Caine 2004: Source waters and flow paths in an alpine catchment, Colorado Front Range, United States. *Water Resources Research* 40, 9. W09401. doi:10.1029/2004WR003076.

Mair, V., A. Zischg, K. Lang, D. Tonidandel, K. Krainer, A. Kellerer-Pirklbauer, P. Deline, P. Schoeneich, E. Cremonese, P. Pogliotti, S. Gruber & L. Böckli 2011: *PermaNET permafrost. Long-term Monitoring Network. Synthesis Report.* INTERPRAEVENT 1, 3. Klagenfurt.

Martin, H.E. & W.B. Whalley 1987: Rock glaciers, part 1: rock glacier morphology: classification and distribution. *Progress in Physical Geography* 11: 260–282.

McClymont, A., J.W. Roy, M. Hayashi, L.R. Bentley, H. Maurer & G. Langston 2011: Investigating groundwater flow paths within proglacial moraine using multiple geophysical methods. *Journal of Hydrology* 399, 1-2: 57–69. doi:10.1016/j.jhydrol.2010.12.036.

Nickus, U., J. Abermann, A. Fischer, K. Krainer, H. Schneider, N. Span & H. Thies (in press): Rock Glacier Äußeres Hochebenkar 1 (Austria) Recent results of a monitoring network. *Zeitschrift für Gletscherkunde und Glazialgeologie.*

Potter, N. 1972: Ice-Cored Rock Glacier, Galena Creek, Northern Absaroka Mountains, Wyoming. *Geological Society of America Bulletin* 83: 3025–3058.

Potter, N., E.J. Steig, D.H. Clark, M.A. Speece, G.M. Clark & A.B. Updike 1998: Galena Creek rock glacier revisited – new observations on an old controversy. *Geografiska Annaler* 80, 3-4: 251–265.

Rogger, M., G.B. Chirico, H. Hausmann, K. Krainer, E. Brückl & G. Blöschl (submitted): Impact of mountain permafrost on flow path and runoff response in a high alpine catchment.

Schmöller, R. & R.K. Fruhwirth 1996: Komplexgeophysikalische Untersuchungen auf dem Dösener Blockgletscher (Hohe Tauern, Österreich). In: Institut für Geographie und Raumforschung der Karl-Franzens-Universität Graz (ed.): *Beiträge zur Permafrostforschung in Österreich.* Grazer Schriften der Geographie und Raumforschung 33: 165–190.

Shroder, J.F., M.P. Bishop, L. Copland & V.F. Sloan 2000: Debris-covered glaciers and rock glaciers in the Nanga Parbat Himalaya, Pakistan. *Geografiska Annaler* 82, 1: 17–31.

Solomon, S., D. Qin, M. Manning, Z. Chen, M. Marquis, K.B. Averyt, M. Tignor & H.L. Miller (eds.): *Climate Change 2007 – The Physical Science Basis. Contribution of Working Group I to the Fourth Assessment Report of the Intergovernmental Panel on Climate Change.* Cambridge, New York.

Stocker, K. 2012: Blockgletscher in Vorarlberg und in der Verwallgruppe. In: Vorarlberger Landesmuseumverein (Hrsg.): *museums verein jahrbuch* 2012: 124–139.

Thies, H., U. Nickus, V. Mair, R. Tessadri, D. Tait, B. Thaler & R. Psenner 2007: Unexpected response of high alpine lake waters to climate warming. *Environmental Science Technology* 41: 7424–7429.

Vitek, J.D. & J.R. Giardino 1987: Rock glaciers: a review of the knowledge base. In: Giardino, J.R., J.F. Shroder & J.D. Vitek (eds.): *Rock Glaciers.* London: 1–26.

Washburn, A.L. 1979: *Geocryology: a survey of periglacial processes and environments. London.*

Wahrhaftig, C. & A. Cox 1959: Rock glaciers in the Alaska Range. *Geological Society of America Bulletin* 70: 383–436.

Whalley, W.B. & H.E. Martin 1992: Rock glaciers: II models and mechanisms. *Progress in Physical Geography* 16, 2: 127–186.

Whalley, W.B., C. Palmer, S. Hamilton & J. Gordon 1994: Ice exposures in rock glaciers. *Journal of Glaciology* 40, 135: 427–429.

Whalley, W.B. & C.F. Palmer 1998: A glacial interpretation for the origin and formation of the Marinet Rock Glacier, Alpes Maritimes, France. *Geografiska Annaler* 80, 3-4: 221–236.

White, S.E. 1971: Rock glacier studies in the Colorado Front Range, 1961 to 1968. *Arctic and Alpine Research* 3, 1: 43–64.

Yershov, E.D. 1998: *General Geocryology. Studies in Polar Research.* Cambridge.

Detecting and Quantifying Area Wide Permafrost Change

Christoph Klug[1], **Erik Bollmann**[1], **Lorenzo Rieg**[1], **Maximilian Sproß**[1], **Rudolf Sailer**[1,2] **& Johann Stötter**[1,2]

[1] *Institute of Geography, University of Innsbruck, Austria*
[2] *alpS – Centre for Climate Change Adaption, Innsbruck, Austria*

1   Introduction

Permafrost and rock glaciers are widespread phenomena in higher altitudes of the European Alps. Since mountain permafrost is very sensitive to the effects of the ongoing climate change (French 1996; Haeberli & Gruber 2009) and its degradation plays a key role in the possible increase of hazard potentials (Kneisel et al. 2007; Noetzli et al. 2007; Sattler et al. 2011), the monitoring of permafrost areas is of particular interest.

Due to the complexity of detecting ground ice, the knowledge regarding permafrost occurence and its distribution is currently very limited for wide areas in the Alps (Kellerer-Pirklbauer 2005; Sattler et al. 2011). The detection and consequent monitoring of possible permafrost areas based on high resolution data (e. g. airborne laser scanning, ALS, and aerial photogrammetry) has the potential to reveal possible permafrost zones which might require further investigations. For an area-wide monitoring of rock glaciers remote sensing techniques have gained in importance in recent years (Kääb 2008).

This research within the permAfrost project focused on three research sites, situated in Tyrol (Stubai and Ötztal Alps) and Vorarlberg (Montafon Range) and ranging from a local to a regional scale.

For detailed investigations based on ALS, four rock glaciers have been chosen for the quantification of surface changes (cf. Fig. 1). These are (1) Reichenkar (RK), (2) Schrankar (SK), (3) Äusseres Hochebenkar (AHK) and (4) Ölgrube rock glacier (OGR). Especially the AHK has a long record of scientific research of nearly 75 years. Different approaches such as terrestrial methods (e. g. geomorphological mapping, geodetic surveys and geological observations), remote sensing (e. g. photogrammetry) and automatic monitoring of different parameters (e. g. ground temperature, climate) have been applied to study the complex rock glacier. Using ALS data in rock glacier research, especially for validating the photogrammetric derived digital elevation models (DEMs), and for calculating flow velocities is an innovative method for area-wide investigations on permafrost.

In a second step, the workflow established on the above mentioned rock glaciers has been adapted to the whole area of Montafon (Vorarlberg). The calculation of surface changes, which has been done on these single rock glaciers, was applied for

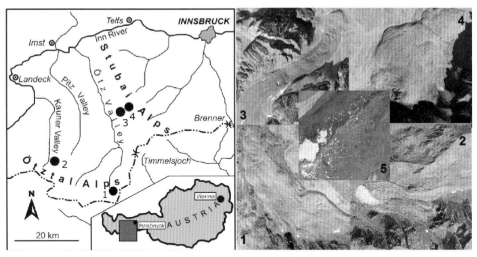

Figure 1: Location of the study sites: Hochebenkar (1), Innere Ölgrube (2), Reichenkar (3), Schrankar (4) and Rofenberg (5). Right side: Orthophotographs from AHK (1), OGR (2) RK (3), SK (4) and Rofenberg with tongue of Hintereisferner (5)

the whole region to detect area wide surface changes. The presented study shows an objective and reproducible approach to assess rock glacier activities (relict or intact) based on two parameters, namely the magnitude and standard deviation of rock glacier thickness changes, calculated from ALS data.

Furthermore, a study site has been installed at Rofenberg for comparing and validating the remote sensing derived insights about permafrost induced surface changes with direct field measurements (e. g. bottom temperature of the snow cover, BTS, temperature loggers, geophysical methods and geomorphologic mapping). This study focused mainly on the demonstration of the possibilities offered by multitemporal ALS data for the detection of local patterns of permafrost distribution in an alpine area.

Therefore the aim of this project was to on i) detect locations where melting of permafrost leads to geomorphologic changes with the use of high resolution data (especially ALS and aerial photogrammetry) and ii) to provide a (operational) methodology for on-going, spatially continuous, area-wide monitoring of geomorphological significant permafrost change.

2   Study sites

2.1   Äußeres Hochebenkar (AHK)

Hochebenkar rock glacier (46° 50' N, 11° 00' E) is a tongue-shaped, talus derived rock glacier about 2 km south of Obergurgl (Ötztal Alps, Austria). It expands from about 2,830 m a. s. l. down to about 2,360 m a. s. l. and reaches a length of 1.1 km at its orographic left, and up to 1.6 km at its orographic right side. A detailed characterisation of the physiognomy of AHK is given by Vietoris (1972) and Haeberli & Patzelt (1982). Since 1938 (Pillewizer 1957) systematic terrestrial investigations on surface flow velocities, front advance rates and surface elevation changes of AHK have been carried out on an annual scale (Schneider & Schneider 2001). Remote sensing methods, like terrestrial and aerial photogrammetry have been applied additionally (Kaufmann & Ladstädter 2002, 2003).

The AHK is characterised by very high surface velocities of up to several meters per year. According to Schneider and Schneider (2001) maximum displacement rates (6.6 m/a) were measured below a terrain edge at about 2,570 m a.s.l. between the 1950/60s. Following Haeberli & Patzelt (1982), these high values are most likely a result of a significant increase of slope steepness below the terrain edge. The related velocity gradients express themselves in deep cross cracks in the rock glacier body (Schneider & Schneider 2001).

2.2   Reichenkar (RK)

Reichenkar rock glacier (47° 02' N, 11° 01' E) is located in the western Stubai Alps (Tyrol, Austria). It is situated in a small, northeast-facing valley called Inneres Reichenkar, which connects to a larger valley, the Sulztal, in the western Stubai Alps. RK is a tongue-shaped, ice cored rock glacier, 1,400 m long with widths that reach 260 m near the head and 170 to 190 m near the front. The rock glacier covers an area of 0.27 km^2 and extends from 2,750 m to an altitude of 2,310 m. The front slope has a steep gradient of 40–41°.

Detailed information about this rock glacier can be found in Krainer & Mostler (2000, 2002), Krainer et al. (2002) and Hausmann et al. (2007). Investigations on surface flow velocities have been carried out in the period 1997 to 2003 with differential Global Positioning System (dGPS) measurements (Chiesi et al. 2003). These data are used to validate the derived horizontal displacement rates. The dGPS measurements were acquired along seven cross profiles at RK with 36 measured points (Krainer & Mostler 2006).

2.3   Schrankar (SK)

The rock glaciers of the Schrankar (47° 03' N, 11° 05' E) are located in the Western Stubai Alps (Tyrol, Austria) in a north to south aligned valley. There are up to 12 rock glaciers of different activities between elevations of 2,400 and 2,800 m a.s.l.. Due to the different activity states, the slope angles of the fronts vary between 32° and 43°. Additionally, beside the rock glaciers, a big dead ice body below the terrain edge of the southern Schrankarferner was identified. Unfortunately, there are no historical surface velocity measurements.

2.4   Ölgrube (OGR)

The rock glacier Ölgrube (46° 53' N, 10° 45' E) is in a small east-west trending site valley of the Kauntertal valley and lies in the western margin of the Ötztaler Alps. Ölgrube consists of two adjacent tongue shaped rock glaciers with different activities and different spatial origins. Although they are separated by a middle moraine, they appear as on rock glacier. It has spatial distribution of 880 m length and 250 width and lies between 2,800 and 2,380 m a. s. l.. The rock glacier has a remarkable front with 60 m height and a slope angle of 38°. The first surface velocity measurements were carried out by Finsterwalder (1928) and Pillewizer (1939), and later on between 2000 and 2005 (Berger et al. 2006; Krainer & Mostler 2006 and Hausmann et al. 2007).

2.5   Rofenberg

The study area Rofenberg (46° 58' N, 10° 48' E) in the Ötztal Alps (Austria) can be characterized as a typical high Alpine environment in mid-latitudes, which ranges between approximately 2,800 and 3,229 m a. s. l.. Areas that were not covered by Little Ice Age (LIA) glaciers have been exposed to permafrost friendly conditions. However, these areas are well suited for investigations aiming the detection of permafrost evidence with a combination of methods (see section 4.4).

2.6   Montafon

The study area Montafon is a 40 km long valley in Austria (Federal State of Vorarlberg), which ranges from the Bielerhöhe to the city of Bludenz to the Silvretta mountain range (see Fig. 13). The valley is surrounded by the Verwall group in the north and the Rätikon and Silvretta group in the south. The highest peak is the Piz Buin (3,312 m a. s. l.) in the Silvretta group. The quantification of vertical and horizontal surface changes in the Montafon Range based on ALS data was one major objective during the last project year. The workflow established on the above mentioned rock glaciers could be successful adapted to the new ALS data from the area of Montafon (Vorarlberg) and the derivation of surface changes, which have been done on these single rock glaciers could be applied for the whole region to detect area wide surface changes.

3   Data

For each of these study sites presented above, basic ALS information is available. These data include at least the coordinates (x, y and z). In exceptional cases additional information about the reflectance of the surface (so called intensity information) are available. Terrain and surface information are separated, resulting in digital terrain models (DTMs), which are significant for the area wide permafrost ALS analysis.

In the year 2006, the relevant part of the Stubai and Ötztal Alps were captured by ALS campaigns, ordered by the Tyrolean Government. The Montafon Range was measured with ALS in 2005 (on behalf of the Government of Vorarlberg). To reach the objectives of the project the introduction of a multi-temporal component in terms of ALS is necessary. For that purpose and in addition to the basic ALS information supplementary ALS information are required. At the Institute of Geography various ALS campaigns have been carried out in the Tyrolean Alps during the last decade. Some of these data are supported to the permAfrost project. ALS data acquisition campaigns at AHK, RK, SK and OGR have been carried out in 2009 and 2010. The 2009 data was acquired within the research project C4AUSTRIA (Climate change consequences on cryosphere, funded by the Austrian Climate Research Programme ACRP), the 2010 data within the MUSICALS project (alps – Centre for Climate Change Adaption Technologies). Additionally, at RK a flight campaign has been carried out in 2007 by the Institute of Geography. The team obtained the permission for the usage of these data in the permAfrost project in order to investigate area-wide geomorphological significant permafrost change. A summary on the exact data acquisition dates is given in Bollman et al. (2011). Table 1 gives an overview of the ALS data, which were used in the permAfrost project. All ALS data have been stored in a Laser Information System (LIS). LIS is able to store and handle ALS point data as well as raster data and is an excellent multi-user environment. Furthermore, the system opens the opportunity to enhance or develop ALS analysis tools.

The ALS campaign for the Montafon Range was organized by the Institute of Geography in the framework of the permAfrost project in autumn 2010. Bad weather conditions (snow fall in higher regions) made it impossible to finish the entire area. Figure 2 gives an overview of the measured flight trajectories in the Montafon

Table 1: ALS data for the analysis of permafrost changes; [1] [2] *entire areas – organized by the Goverments of Tyrol and Vorarlberg;* [3] *Reichenkar;* [4] [5] *Reichenkar, Schrankar, Hochebenkar, Ölgrube;* [6] *permAfrost ALS flight campaign Montafon;* [7] *Reichenkar (not used 2011)*

Region	2005/2006	2007	2009	2010	2011
Stubai Alps	x [1]	x [3]	x [4]	x [5]	x [7]
Ötztal Alps	x [1]		x [4]	x [5]	x [7]
Montafon Range	x [2]			x [6]	

Table 2: Minimum point density

Region	2005 / 2006	2007	2009	2010	2011
Stubai Alps	1 pt / 4 m^2	2 pts / m^2	2 pts / m^2	2 pts / m^2	2 pts / m^2
Ötztal Alps	1 pt / 4 m^2		2 pts / m^2	2 pts / m^2	3 pts / m^2
Montafon Range	1 pt / m^2			2 pts / m^2	4 pts / m^2

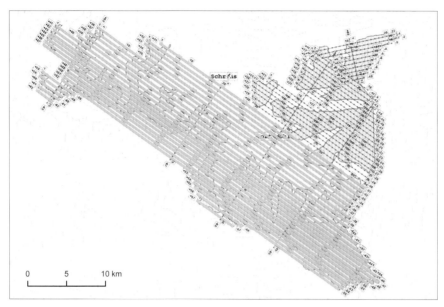

Figure 2: Flight plan and trajectories acquired in autumn 2010 (blue) and in autumn 2011 (red), Montafon Range

Range in 2010 (blue). The northern part has been measured in autumn 2011. Compared to the point density of the first flight campaign in 2010, the laser scanning data has been ordered with a point density of at least 4 pts/m² (cf. Table 2).

The higher point density and the expansion of the flight area were enabled by an agreement between the Institute of Geography, University of Innsbruck and the Landesvermessungsamt Vorarlberg (department VoGIS).

Additionally to the flight campaign of the Montafon, on 6 November 2011 a GPS campaign was carried out to support the company TopScan with GPS reference data, which will be used to georeference the ALS raw data. 72 GPS points were measured in the south-eastern part of the Montafon Range close to Silvretta lake (Bieler Höhe, Silvretta group) with an average accuracy (x, y and z) of approximately 10 to 17 cm.

3.1   ALS data quality control and data processing

In general the data point cloud data (x, y, z and intensity). As mentioned above, data are stored in the LIS data base. In case of permAfrost a filtering of first or last pulse was, due to the absence of higher vegetation (trees, scrubs) in permafrost terrain, not necessary.

The ALS data quality control is an important step in the ALS based permafrost analysis work flow. The quality control of the archive (2005/2006, 2007, 2009 and 2010) data has been done within the framework of relevant projects (e. g. C4AUS-TRIA, MUSICALS). The quality control of the ALS campaign Montafon has been

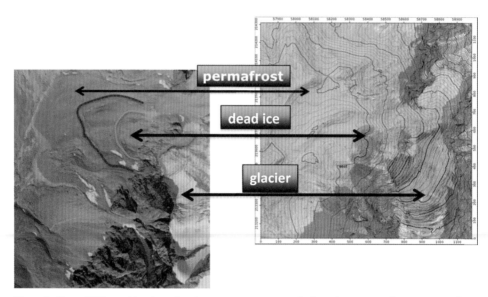

Figure 3: Part of SK, combination of various remote sensing methods and mapping of various cryospheric processes (glacier dynamic, dead ice development and permafrost) calculated by differencing ALS data

done immediately after data delivery. Particular efforts have been undertaken to improve existing data sets and in particular to check the quality of the Montafon ALS data. The first part of the study aimed at quantifying the vertical accuracy of the available multi-temporal ALS data sets. In comparison with differential dGPS measurements, the vertical accuracy of the ALS point data is in the range of ±0.10 m (mean absolute error), with a standard deviation of ±0.10 m for relatively flat and homogeneous areas. Including point to raster aggregation errors as well, the cumulative errors (e. g. <0.12 cm in 40° steep and rough terrain) are also promising. Hence, surface elevation changes which exceed these error margins can be assigned to permafrost related activities.

On base of these point data DTMs were calculated using different algorithms and settings. Good results were achieved with cell sizes down to 1 m, or even down to 0.5 m, if gaps were closed. DTM from different years have been used to analyse volume changes and to create a surface changes map by simply differencing the DTMs. It is obvious that ALS is able to deliver information about terrain and changes therein where, due to shadow effects, no information are available in orthoimages (Fig. 3).

Figure 3 gives an impression of how data from two ALS campaigns are combined to get information about altitudinal changes in an Alpine terrain. It captures the entire Schrankar area and shows the altitudinal changes corresponding to cryospheric processes between 2006 and 2009. Large changes can be observed in glaciated areas, moderate changes in dead ice areas and where debris flows occurred and small changes where permafrost melting is assumed. The know-how for the data handling, the quality control and the co-registration have been acquired at the Institute of Geography during various ALS related projects (e. g. C4AUSTRIA (Climate Change Consequences on the Cryosphere), Austrian Climate Research Program), which lead to synergistic effects (application of already developed tools or improvement of existing tools for example) across the linked projects.

4 Project related activities

4.1 Quantifying surface changes on rock glaciers: Photogrammetric and ALS measurements

The quantification of vertical and horizontal surface changes based on ALS data was one major objective within the second project year. These calculations are useful in order to get more exact information about the response of permafrost phenomena to the process of climate change and build an important analytical component for permafrost distribution mapping, because it allows identifying the state of activity of rock glaciers. Within this objective we reconstructed and analysed the surface displacements on four rock glaciers in the Stubai and Ötztal Alps. These results are derived from aerial images and ALS data over the period 1953/54 to 2011. Using ALS data in rock glacier research, especially for validating the photogrammetric derived DEMs, and for calculating flow velocities is an innovative method for area-wide investigations on permafrost.

4.1.1 Data

The basic data applied in the analysis are on the one hand greyscale aerial photographs, provided by the Austrian Federal Office of Metrology and Survey (BEV), with a scale from 1:16,000 to 1:30,000. For the study sites, analogue aerial stereoscopic pairs were available for the years 1953/54, 1969, 1971, 1973, 1977, 1989, 1990, 1994 and 1997, taken with Wild/Leica-Cameras of different types. Additionally, the federal province of Tyrol provided high resolution orthoimages from the years 2003 and 2009.

To pursue the objective target at best, the date of recording is very important, because the objects of interest are situated at around 2,300–2,900 m a. s. l, and snow-covered areas would make DEM generation impossible. Another problem concerning DEM generation is the topographic shadowing effects. For that reason, only aerial photographs with a certain standard were used.

On the other hand ALS data were applied, which have been acquired in flight campaigns at AHK, RK, SK and OGR in 2006, 2009, 2010 and 2011. The 2006 campaign was initiated by the federal province of Tyrol. As mentioned above, the 2009 and 2011 data were acquired within the C4AUSTRIA, the 2010 data within the MUSICALS project. DEMs out of these ALS flight campaigns were generated and integrated in the analysis. Additionally, at RK a flight campaign has been carried out in 2007.

4.1.2   Method

For the photogrammetric analyses the analogue data were provided by BEV as scanned aerial photographs (ca. 1,600 dpi). Additionally the required calibration reports were supplied. Image orientation, automatic DEM extraction and digital orthophoto generation were performed within the application LPS of the software ERDAS, as well as the application OrthoEngine of the software Geomatica.

For detailed description of the method see Baltsavias (2001), Kääb (2004) and Kraus (2004). The DEMs were generated from the mono-temporal stereo-models with 1 m spacing. Additionally, the orthoimages were calculated with the resampling-method cubic convolution and with 0.2 m ground resolution.

Afterwards an accuracy assessment is performed by comparing the generated DEM, especially with the high-quality DEM of the ALS flight campaign of 2009. Therefore three areas are detected, which represent the surface structure (slope, aspect, height) of the rock glaciers but are located in stable regions. On the basis of this analysis the root mean square error (RMSE) could be calculated. The computed RMSE for AHK, RK, SK and OGR (cf. Table 3) show values of 0.2–1.3 m. Afterwards the vertical changes were derived through subtraction of the multitemporal DEMs over the different periods.

The horizontal displacement rates out of the orthoimages were calculated with the Correlation Image Analysis (CIAS) Software (Kääb & Vollmer 2000). CIAS identifies displacement rates as a double cross-correlation function based on the grey values of the used input images (Kääb 2010). The correlation algorithm searches via block matching a predefined corresponding reference section in the image of acquisition time 1 (t_1) in a sub-area of image of acquisition time 2 (t_2). For measuring changes in

Table 3: Calculated RMSE for the generated DEM vs. ALS DEM 2009. (–) Aerial images not available or acquisition in winter

DEM	RK-RMSE	AHK-RMSE	SK-RMSE	OGR-RMSE
2011 (ALS)	0.10 m	0.11 m	0.12 m	0.09 m
2010 (ALS)	0.11 m	0.12 m	0.11 m	0.13 m
2009 (ref. ALS)	ref	ref	ref	ref
2006 (ALS)	0.13 m	0.10 m	0.12 m	0.11 m
1997	–	0.41 m	–	
1994	0.22 m	–	0.32 m	
1989 /1990	0.26 m	0.42 m		
1977	–	0.25 m	–	–
1973	1.29 m	–	0.64 m	0.83 m
1971	–	0.55 m		–
1969	–	0.29 m	0.42 m	–
1953 / 1954	0.78 m	0.58 m	0.63 m	0.55 m

geometry, relative accuracy is more important than the absolute position of the images (Kääb 2002). Therefore, the multitemporal orthoimages have been co-registered using tie points in addition to the initial georeferencing.

Additionally the horizontal displacement rates are calculated from shaded reliefs of ALS DEM with the open source image correlation software Imcorr (Scambos et al. 1992), which applies also digital matching techniques using normalized cross-correlation (NCC) to calculate flow velocities (Scambos et al. 1999).

The accuracy of the calculated horizontal displacement rates on AHK based on ALS DEM is evaluated using dGPS data. As the dGPS measurements and ALS data acquisitions have not been carried out at the same dates, the ALS based displacement rates are adjusted to the time span between the dGPS data acquisitions using

$$\overline{dALS}_{t1,t2} = ALS_{t1,t2} / \Delta tALS_{t1,t2} \cdot \Delta tdGPS_{t1,t2} \tag{1}$$

where $\overline{dALS}_{t1,t2}$ is the adjusted displacement rate [m / a] of the Imcorr results, $ALS_{t1,t2}$ is the displacement rate of the original Imcorr results, $\Delta tALS_{t1,t2}$ is the number of days between the ALS flight campaigns and $\Delta tdGPS_{t1,t2}$ is the number of days between the dGPS measurements (Bollmann et al. 2012). The time adjusted Imcorr output point file as well as the interpolated displacement raster values are compared to the dGPS displacement rates. For the generation of displacement grids a trimmed mean interpolation function (search radius 2 m) was used. This allowed the removal of erroneous measurements (outliers) (cf. Bollmann et al. 2012). For the 2009 / 2010 time period 39 dGPS measurements from the cross profiles *L0* to *L3* are used for validation. For the period 2006 / 2010 dGNSS data from *L1* is not available, therefore

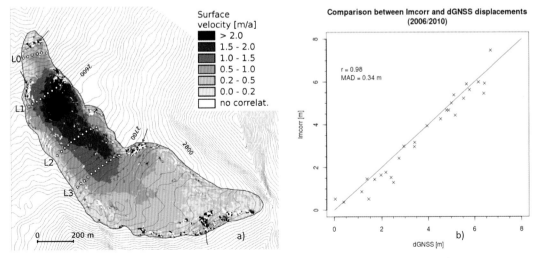

Figure 4: a) Mean annual surface velocity of AHK between 2006 and 2010. Results interpolated from Imcorr output. L0 to L3 (white circles) indicate location of dGPS measurement points along 4 cross profiles. Arrows indicate areas with artifacts. b) Comparison between Imcorr and dGPS displacements on AHK between 2006 and 2010

Table 4: Comparison between horizontal displacement rates calculated from CIAS (orthoimages) and Imcorr (ALS-data) with dGPS (Imcorr – dGPS) for the periods 1997–2003, 2006–2010 and 2009/2010. In columns with "Pts." to original CIAS/Imcorr output points are compared to dGPS, otherwise the interpolated rasters

CIAS / Imcorr – dGPS	mean [m]	abs.mean [m]	std [m]	max [m]	min [m]	RMSE [m]	R²
OP CIAS Pts. 1997-2003	0.24	0.4	0.46	1.28	0.02	2.05	0.91
ALS Imcorr Pts. 09-10	0.09	0.28	0.50	2.50	0.66	0.50	–
Raster 09-10	0.06	0.27	0.48	2.42	0.69	0.47	0.98*
Pts. 06-10	0.50	0.68	3.57	18.6	1.02	3.54	–
Raster 06-10	-0.25	0.28	0.44	0.80	1.22	0.50	0.94*

only 28 reference measurements could be used (see Fig. 4a, b). The vertical accuracy of the generated DTMs at AHK is determined using eleven dGPS measurement sites. The eleven fix points are not influenced by any topographic changes.

Furthermore the results of RK based on orthoimages were compared with measured dGPS flow velocities. The original CIAS output point files, as well as the automatically measured displacement vector values, were compared to the dGPS displacement rates, using a buffer of 3 m around the dGPS points. Table 4 presents the calculated differences by CIAS and Imcorr, considering the mean absolute error (*abs. mean*) in combination with the standard deviation (*std*) to be the most appropriate accuracy indicators.

For the time period 1997–2003, an absolute mean deviation of 0.4 m (*std* 0.46 m) between the dGPS values and the corresponding nearest CIAS point measurements was calculated. Figure 4b shows the computed and the 36 dGPS measured flow velocities within the investigated periods (cf. Fig. 5a, b). Additionally, the dGPS points were used as reference for the periods 1954–1973 and 2003–2009. In contrast to the latter epoch, the velocities between 1954 and 1973 were slower in evidence, but there is still a good relative correlation within the profiles.

4.1.3   Results and Interpretation

In the case of RK, AHK and OGR permafrost creep is the most important factor governing surface elevation changes. Additionally, the derived high annual mean velocities on RK revealed further processes, which could be involved in surface displacement. Horizontal and vertical displacements of RK, AHK, SK and OGR could be successfully calculated from multitemporal aerial photographs and ALS data over a period of 57 years. The existing knowledge about the flow behavior of the three rock glaciers could be confirmed with the method and reconstructed. Furthermore the length of the time series allowed the investigation of temporal variations in flow velocities, which has little been done for a fast flowing type like RK or AHK. The study revealed an acceleration of the investigated rock glaciers since the late 1990s (cf. Table 5). In the following the results of RK and AHK will be presented in detail, whereas some results of OGR and SK will be presented as figures (cf. Fig. 9 and 10).

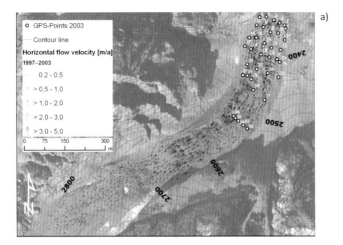

a)

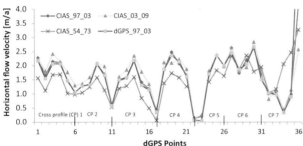

b)

Figure 5: a) Mean annual horizontal flow velocity on RK between 1954–1973. White circles indicate location of dGPS measurement points along 7 cross profiles. b) Comparison of the measured (dGPS) and computed (CIAS) flow velocities for periods 1997–2003, 2003–2009 and 1954–1973

In contrast to AHK, where the displacement rates have been similar to nowadays in the 1950s and 1960s, the RK shows acceleration over all periods from 0.75 to 1.3 m/a nowadays. Beside the temporal variation, also a spatial variation in the velocity field of RK could be detected. So it seems that the orographic left part of the rock glacier in the middle and lower section show higher velocities.

4.1.3.1 Reichenkar

The horizontal surface velocities and the thickness changes on RK (cf. Fig. 6) varied between the individual investigated epochs. From 1954–2011 an average vertical loss of the rock glacier of nearly 3.2 m (ca. –6 cm/a) can be observed. In the rooting zone surface lowering of up to –0.5 m/a in various periods indicates massive loss of ice. The average elevation change in this zone lies within the range of –0.10 m/a.

The increase in thickness at the front of individual flow lobes suggested that elevation changes are influenced by mass advection. Over the investigated periods, the surface of RK was creeping with average rates of 0.9 m/a (cf. Table 5). Maximum creep rates occur in a transition zone (2,540–2,660 m a.s.l.), a steeper mid-section before the flat tongue, where compressive flow with development of transverse ridges and furrows can be observed. The measured surface displacements depict acceleration since the late 1990s, from the middle to the front part of the rock glacier. The

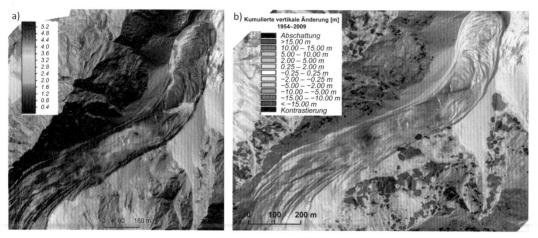

Figure 6: a) Displacement field (unclassified) of RK derived from generated DEM based on ALS data (time 1: 30 September 2009, time 2: 7 October 2010). Ground resolution was 50 cm. 6736 homologous points were matched on RK. b) Difference image of RK, calculated by differencing DEM of 1954 generated from aerial images and ALS data from 2009. Black and purple areas indicate errors because of bad contrast or shadowing effects.

displacement rates in the root zone seem to be constant over the whole period (0.2–0.5 m/a), whereas flow rates at the tongue, which were consistent over the periods from 1954–1997 with average rates of 1–2 m/a along the central axis, increased to 2–3 m/a since 1997. The increasing velocities cannot be explained by the gradient of the slope, which measures only 11–12° on the lower part of RK.

The front of RK advanced 53 m from 1954 to 2009 (0.76 m/a). Figure 7 depicts the mentioned acceleration, which increased from an average of 0.6–1.55 m/a during the last 20 years. The measurements show highest displacement rates at the front

Table 5: Calculated horizontal velocity from orthoimages with CIAS (mean, maximum velocity & acceleration dv) of all periods on RK and AHK. The two rock glaciers were separated into three sections (root zone, transition zone and tongue), in every sector the same amount of measured blocks was used for calculation of the annual mean (cf. Klug et al. 2012)

Period	Measured blocks	Mean velocity [m/a]	Max. velocity [m/a]	dv [%]
Reichenkar				
1954–1973	841	0.75	3.4	–
1973–1989	871	0.73	3.5	–2.7
1989–1994	909	0.73	4.2	0.0
1994–1997	987	0.76	5	+ 3.2
1997–2003	830	1.12	4.9	+ 47.4
2003–2009	811	1.3	4.1	+ 16.1
Hochebenkar				
1953–1969	2438	0.84	5.2	–

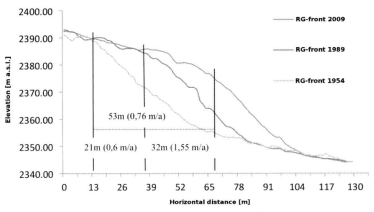

Figure 7: Front advance depicted by longitudinal profiles

within the range of up to 3 m / a. At the orographic left part of the flat tongue (2,520–2,460 m a. s.l.), one part features relatively high movement rates (ca. 3.3 m / a) since 1994. In combination with the general topography the distinct velocity gradients indicate that the lower part of the frozen body is overridden from above by a new lobe (Klug et al. 2012).

On the base of the calculated vertical changes and the different delineated areas it was tried to derive the balance of the rock glacier RK and to calculate the water equivalent. The difference grids have been separated in positive and negative values, whereat areas representing errors caused by bad contrast or shadowing effects in photogrammetrical derived differences have been excluded. The mean annual *dz* integrated over the rock glacier area amounts to –5.6 cm / a. By multiplying the calculated *dz* with the density of ice (900 kg m^{-3}) the water equivalent has been calculated, which amounts to a mean annual loss of –58.3 mm w. e. (cf. Table 6).

Table 6: Mass balances on RK

Period	Mass gain [m³]	Mass loss [m³]	Balance [m³]	m³ / a	dz / a [m]	mm w. e. / a
1954–2011	209,908	1,188,932	−979,024	−17,175.9	−0.064	−58.3
1954–1973	60,403	348,471	−288,068	−16,945.2	−0.063	−51.5
1973–1989	58,167	331,322	−273,155	−15,175.3	−0.056	−58.0
1989–1994	23,856	88,347	−64,491	−12,898.2	−0.048	−43.8
1994–2006	44,309	247,332	−203,023	−16,918.6	−0.063	−57.5
2006–2009	34,945	80,848	−45,903	−15,301.0	−0.057	−52.0
2006–2007	24,396	33,922	−9,526	−9,526.0	−0.035	−32.4
2007–2009	28,259	62,181	−33,922	−16,961.0	−0.063	−57.6
2009–2010	33,470	50,728	−17,258	−17,258.0	−0.064	−58.6
2010–2011	27,353	42,496	−15,143	−15,143.0	−0.056	−51.4

4.1.3.2 Hochebenkar

The rock glacier is characterized by a comparatively high flow velocity of several me-
tres per year and periodically changing flow rates between 1953 and 2011 (cf. Table 5). A transverse terrain edge at an altitude of about 2,580 m a. s. l. divides the rock
glacier into a lower steep part and an upper flat part. Within all periods, the rooting
zone shows constant velocities (0.2–0.5 m / a), whereas in the adjacent zone towards
the terrain edge displacement rates differ during the investigation periods.

In the period from 1953 to 1969, high movement rates (1.0–2.5 m / a) were measured in this zone. At the same time, the upper part of the tongue showed the development of massive transverse cracks, where the loss of mass was extremely high.
From the early 1970s to the beginning 1990s, a phase with relatively slow average
annual velocity rates (0.5–1.0 m / a, Table 3) could be shown. From the 1990s on-
wards, the movement rates increased, showing velocities (1.0 to 2.5 m / a) similar
to that of the late 1960s (Fig. 8). On the terrain edge, creep velocities increased to
about 2.5 m / a. Maximum creep rates (6.9 m / a) have been measured in this transition zone from 1953 to 1969, although rockfall may be a significant process here and
cannot be attributed to permafrost creep exactly.

Surface elevation change rates > 5 m are identified in several areas of AHK. At the
orographic right margin of the rock glacier values > 0.5 m are caused by creep processes that resulted in a downward transport of mass and expansion of AHK between
2006 and 2010 (Fig. 11). Distinct elongate-shaped alterations of positive and negative values occur on the main part of the rock glacier, especially between 2,730 and
2,820 m at the orographic right side. They clearly result from advancing ridges and
furrows. Between *L0* and *L1*, at about 2,500 m at the middle part of AHK, a sharp

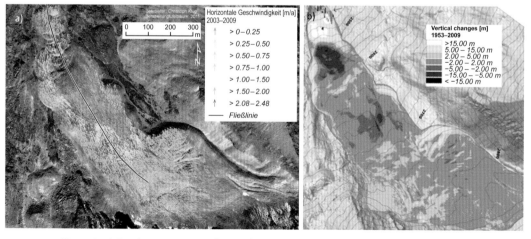

*Figure 8: a) Displacement vectors of AHK derived from generated orthoimages (time 1: 9 September 2003,
time 2: 18 September 2009). Ground resolution was 20 cm. 2,439 homologous blocks were successfully
matched using CIAS. b) Cumulated vertical changes of AHK between 1953 and 2009. Arrow at the front
indicates errors because of bad contrast*

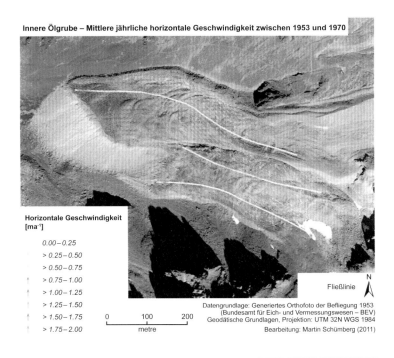

Innere Ölgrube – Mittlere jährliche horizontale Geschwindigkeit zwischen 1953 und 1970

Horizontale Geschwindigkeit
[ma⁻¹]

0.00 – 0.25
> 0.25 – 0.50
> 0.50 – 0.75
> 0.75 – 1.00
> 1.00 – 1.25
> 1.25 – 1.50
> 1.50 – 1.75
> 1.75 – 2.00

Fließlinie   N

0   100   200
        metre

Datengrundlage: Generiertes Orthofoto der Befliegung 1953
(Bundesamt für Eich- und Vermessungswesen – BEV)
Geodätische Grundlagen, Projektion: UTM 32N WGS 1984

Bearbeitung: Martin Schümberg (2011)

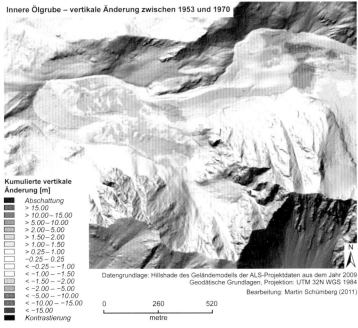

Innere Ölgrube – vertikale Änderung zwischen 1953 und 1970

Kumulierte vertikale
Änderung [m]

Abschattung
> 15.00
> 10.00 – 15.00
> 5.00 – 10.00
> 2.00 – 5.00
> 1.50 – 2.00
> 1.00 – 1.50
> 0.25 – 1.00
–0.25 – 0.25
< –0.25 – –1.00
< –1.00 – –1.50
< –1.50 – –2.00
< –2.00 – –5.00
< –5.00 – –10.00
< –10.00 – –15.00
< –15.00
Kontrastierung

N

Datengrundlage: Hillshade des Geländemodells der ALS-Projektdaten aus dem Jahr 2009
Geodätische Grundlagen, Projektion: UTM 32N WGS 1984

Bearbeitung: Martin Schümberg (2011)

0   260   520
        metre

Figure 9: above – Displacement vectors of OGR derived from generated orthoimages (time 1: September, 1953, time 2: 29 September 1970). Ground resolution was 20 cm. 2,149 homologous blocks were successfully matched using CIAS. below – Cumulated vertical changes of OGR between 1953 and 1970

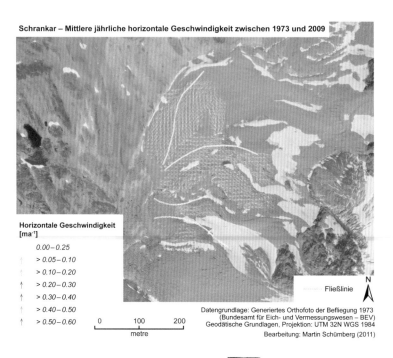

Schrankar – Mittlere jährliche horizontale Geschwindigkeit zwischen 1973 und 2009

Horizontale Geschwindigkeit [ma⁻¹]

$0.00-0.25$

$> 0.05-0.10$

$> 0.10-0.20$

$> 0.20-0.30$

$> 0.30-0.40$

$> 0.40-0.50$

$> 0.50-0.60$

Fließlinie

N

0   100   200
metre

Datengrundlage: Generiertes Orthofoto der Befliegung 1973
(Bundesamt für Eich- und Vermessungswesen – BEV)
Geodätische Grundlagen, Projektion: UTM 32N WGS 1984
Bearbeitung: Martin Schümberg (2011)

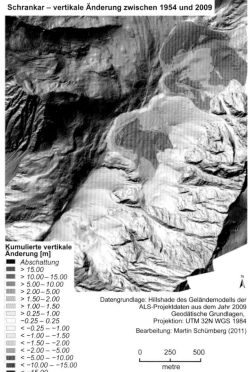

Schrankar – vertikale Änderung zwischen 1954 und 2009

Kumulierte vertikale Änderung [m]

Abschattung
> 15.00
$> 10.00-15.00$
$> 5.00-10.00$
$> 2.00-5.00$
$> 1.50-2.00$
$> 1.00-1.50$
$> 0.25-1.00$
$-0.25-0.25$
$< -0.25--1.00$
$< -1.00--1.50$
$< -1.50--2.00$
$< -2.00--5.00$
$< -5.00--10.00$
$< -10.00--15.00$
< -15.00
Kontrastierung

N

Datengrundlage: Hillshade des Geländemodells der
ALS-Projektdaten aus dem Jahr 2009
Geodätische Grundlagen,
Projektion: UTM 32N WGS 1984

Bearbeitung: Martin Schümberg (2011)

0   250   500
metre

Figure 10: above – Example of displacement vectors of SK derived from generated orthoimages (time 1: September, 1973, time 2: 29 September 2009). Ground resolution was 20 cm. 2,149 homologous blocks were successfully matched using CIAS. Below – Cumulated vertical changes of SK between 1954 and 2009

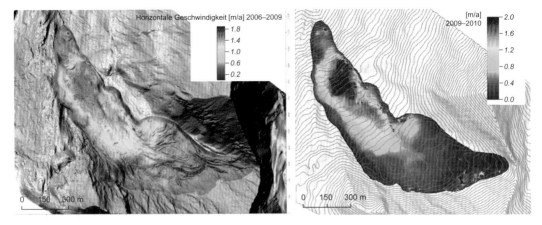

Figure 11: Displacement field (unclassified) of AHK derived from generated DEM based on ALS data (time 1: 18 September 2006; time 2: 30 September 2009). Ground resolution was 50 cm. 7,751 homologous points were successfully matched using Imcorr (left). Displacement field (unclassified) of AHK (time 1: 30 September 2009, time 2: 7 October 2010). Ground resolution was 50 cm. 7,912 homologous points were successfully matched using Imcorr on AHK (right)

transition from areas with high surface elevation decrease and gain is evident. These areas correspond well with the area described by Haeberli & Patzelt (1982) and Schneider & Schneider (2001) where deep cross cracks occur as a consequence of local sliding of the rock glacier on the bedrock and increased tension in the permafrost body.

In the upper part the measurements show elevation changes within the range of –0.10 m / a, whereas the changes increase to nearly –0.6 m / a at the zone below the terrain edge (Fig. 12). Thinning of the frozen debris at the terrain edge of around 2,580 m a. s. l. is compensated to a large extent by corresponding thickening in the lowest part. Below 2,580 m, which marks the end of the steady-state creeping zone, the tongue has moved into very steep terrain. In this area landslides have occurred due to the specific topographic situation and make image correlation nearly impossible. The front of AHK advanced from 1953 to 2011 135 m (2.4 m / a). Since 1969 the rates of front advance decreased from 4.1 m / a to 1 m / a in the 1990s, since that time an increase could be observed to 1.6 m / a nowadays.

In general, the detected local surface elevation changes of AHK are caused by horizontal displacements of the creeping rock glacier. Regarding the whole rock glacier, an area-averaged mass loss is not detected. Therefore, the ice content of AHK seems to be well protected from surface energy input and significant ice melt did not occur between the investigated periods.

In a detailed investigation of the ALS data 2006, 2009 and 2010 of AHK, individual surface velocity fields have been detected for the time period 2006 to 2010 (Fig. 11). In general, the mean annual velocity increases gradually from < 0.2 m / a at the root zone to a maximum of > 2 m / a in the middle part of the rock glacier and at its orographic right side at about 2,600 m a. s. l..

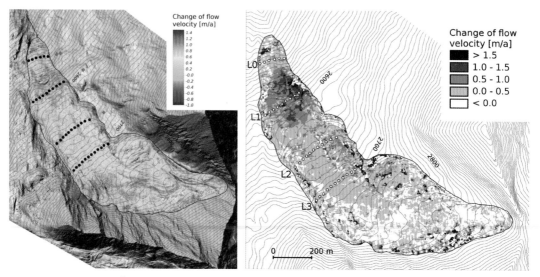

Figure 12: Changes in mean annual flow velocity unclassified (above) and classified (below) between the time period 2006/2009 and 2009/2010

Maximum velocities correspond well with an increase of terrain steepness below 2,640 m. Below 2,560 m the surface velocity decreases again, even though the terrain is still steep. A complete decoupling of the upper and lower part of AHK, as has been assumed by Haeberli & Patzelt (1982), cannot be found in the analysis of the 2006 to 2010 velocity fields. The velocity decrease between *L1* to *L0* is rather gradually than abrupt. However, the lowest part of AHK (around *L0*) is clearly influenced by other creep characteristics than the part above 2,560 m (Fig. 12).

For some areas no creep rates could be calculated, because no correlations between the two input images are made by Imcorr. These areas either occur at steep lateral sides or in fast creep steep areas. In these areas the surface topography changes strongly due to rotation of boulders and surface instabilities. Several velocity artifacts (arrows in Fig. 8), resulting from miss-matching of the two Imcorr input images occur. In most cases, they are spatially connected to areas where no correlation of the two images could be made. Such areas have to be excluded from interpretation.

Between the period 2006/2009 and 2009/2010 a general velocity increase is detected. Comparing Figure 12 left and right shows, that the absolute increase of surface velocity is highest for areas that initially show high mean annual velocities. The most significant acceleration in the order of 1.0–1.5 m/a occurs in the steep part of the rock glacier at the cross profile line *L1*. Velocity changes in that area have been discussed by Haeberli & Patzelt (1982) and Schneider & Schneider (2001).

Most likely, they do not present variations in internal creep characteristics, but are a result of sliding at the base of the creeping permafrost body on the bedrock. Over large parts of the rock glacier, a velocity increase in the order of 0.0–0.5 m/a is observed, whereas towards the root zone slightly negative values are calculated. These values are close or below to the level of significance of 0.3 m (cf. Table 4; *std* raster 09 / 10). Thus, considering the accuracy of the data and method, it cannot be state that the velocity really decreased in the root zone of AHK.

4.1.4   Conclusion

Although it is impossible to derive the flow mechanics of a rock glacier from its measured velocities, some conclusions may be drawn out of the area-wide measured velocity field. Some regional-scale studies have been conducted (cf. Roer 2005), but most investigations are still concentrated on single rock glaciers and mesoscale information on this topic is limited. Area-wide data about the flow behaviour of rock glaciers provide useful input data for modelling rock glacier dynamics. From the calculated multi-temporal displacement rates conclusions can be drawn on driving factors of rock glacier creep.

Horizontal displacements of the investigated rock glaciers could be calculated from multitemporal aerial images and ALS data. Comparison between the surface displacement rasters and dGPS data indicate an accuracy (standard deviation) of the calculated displacement rates of 0.3 m for the period 2009 / 2010 respectively 0.5 m (0.13 for an annual scale) for the period 2006 / 2010. As AHK is a very fast creeping rock glacier, the data and method used is sufficient to receive significant results for almost all parts of the rock glacier. However, a time span smaller five years between the ALS data acquisition data and raster resolution of 0.5 m might not be long enough to obtain significant results for slow creeping rock glaciers.

The developed workflow could be easy adapted to the new ALS data from the area of Montafon (Vorarlberg) and the derivation of surface changes, which have been done on these single rock glaciers could be applied for the whole region to detect area wide surface changes. The Figures 3, 4 and 5 show results, which are derived from the orthoimages and ALS flight campaigns 2006 and 2009. They demonstrate the horizontal flow velocities between different periods. Table 6 shows the computed flow velocities from the orthoimages of AHK and RK.

In the case of AHK, permafrost creep is the most important factor governing local surface elevation changes. Future applications of ALS in rock glacier monitoring should focus on (I) the method's capability to quantify very slow creeping permafrost, (II) the performance on rock glaciers with small changes of local surface topography (smooth surface), (III) on the calculation of 3D displacements, and (IV) and area-wide quantification of rock glacier creep.

In the case of RK and AHK permafrost creep is the most important factor governing surface elevation changes. Computer-based aerial photogrammetry in combination with ALS allows detailed determination and analysis of surface elevation changes in and horizontal displacements on the investigated rock glaciers over a period of 56 years. These calculations build an important analytical component for identifying the state of activity of rock glaciers.

Using ALS data in rock glacier research, especially for validating the photogrammetrical derived DEMs, and for calculating flow velocities has proven to be a useful

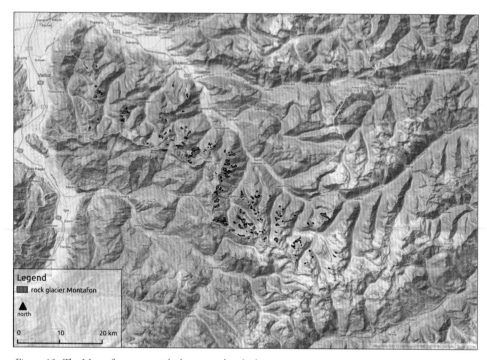

Figure 13: The Montafon range with the mapped rock glacier inventory

data source for area-wide investigations on rock glaciers. The results show the potential of the method combination to quantify spatio-temporal variations of rock glacier surface changes.

4.2   Rock glacier inventory of the Montafon based on ALS data

4.2.1   Data

As mentioned in section 3, the ALS campaign of the Montafon Range was organized by the Institute of Geography in the framework of the permAfrost project in autumn 2010. Due to bad weather conditions (snow fall in higher regions) the entire area could only be finished after a second campaign in autumn 2011. The missing ALS data of the Montafon Range have been delivered in autumn 2012. Figure 2 gives an overview of the measured flight trajectories in the Montafon Range. The ALS data were preprocessed by the responsible companies, which includes the determination of the absolute position and orientation of the laser scanning system during the flight, as well as the system calibration and strip adjustments (cf. Wehr & Lohr 1999). After preprocessing, the delivered last pulse ALS point clouds consisting of x, y and z coordinates were imported and stored in our laser data information system. Compared to the point density of the first flight campaign in 2010, the laser scanning data has been ordered with a point density of at least 4 pts / per m² (cf. Table 2).

ALS data quality control, data processing and its management process is analogue to that described in section 3. On the base of 72 measured dGPS points an average slope dependent vertical accuracy of approximately 14–35 cm could be calculated.

4.2.2 Method

The rock glacier inventory of the Montafon area was established based on field work, geomorphological maps and ortho-images. For the compilation of the inventory, we identified and delineated rock glaciers based on their specific morphological appearance (Barsch 1996) using colour orthophotos and laser scan images. Following Krainer & Ribis (2012), most rock glaciers could be easily recognized due to their typical morphological features. However, the definition of the upper boundary in the rooting zone was difficult and to some extent arbitrary. A few rock glaciers were difficult to recognize at all as their morphological outlines are not clearly defined. In these cases the surface morphology (indication downslope creep) was the criterion for definition as rock glacier. As a transition exist between large solifluction lobes and small rock glaciers, the size of a rock glaciers was defined by a minimum length of 50 m and a minimum width of 35 m. Altogether 193 rock glaciers were detected.

As rock glaciers have a highly complex surface topography, and furthermore the quality of our analysis strongly relies on an accurate terrain representation, we generated DTMs with 1 m x 1 m cell size for each rock glacier and its near surrounding (buffer of 200 m around the rock glacier outline). To do so, a mean function contained in SAGA GIS, was applied to generate a DTM of 2004 as well as of 2010 of each rock glacier. The mean function populates each grid cell with the average elevation values of all ALS points that are spatially contained within each cell. As the given point densities do not allow to populate each 1 m x 1 m DTM cell with elevation values directly, a close gaps tool was applied to populate cells that did not originally contain elevation values.

For the conduction of our study, we used the same rock glacier extents as derived in the compilation of the rock glacier inventory. Each rock glacier outline is defined by a polygon stored in shapefile. The usage of multi-temporal ALS datasets allowed the identification of terrain surface changes. Geometric changes of rock glaciers caused by melt, creep or material input are quantified by means of DTM differencing. For that purpose, thickness change $dDTM^n$ is calculated for each of the 193 rock glaciers ($n = 1$–193) by subtracting the 2004 DTM from the 2010 DTM with

$$dDTM^n = DTM^n - DTM^n \tag{2}$$

where $dDTM^n$ is the thickness change model of rock glacier n between 2004 and 2010, furthermore the area-averaged thickness change of each rock glacier was calculated. A first classification of the rock glaciers was done on the base of these ALS differences between 2004 and 2010.

4.2.3 Results and interpretation

In our study we assumed, that rock glacier thinning is caused by melt of internal ice followed by surface subsidence. Thus, thinning can only occur at intact rock glaciers. Regarding the spatial distribution and magnitude of thickness change on a whole rock glacier, the magnitude of thinning taking place in one part of the rock glacier might be balanced by frost heave or debris input in another part of the rock glacier. This would result in an area-averaged thickness change $(\overline{\Delta z^n})$ of 0 m leading to the erroneous conclusion that the rock glacier has to be relict. To overcome this problem of misinterpretation, the standard deviation σz^n was calculated, because it contains additional information about the spatial variability of the rock glacier deformation.

A large σz^n is considered to be representative either for an inactive rock glacier where different magnituds of thickness change occur at different areas on the rock glacier or for an active rock glacier where big σz^n is caused by the displacement of individual furrows and ridges of the rock glacier. In any case, rock glacier with big σz^n are considered to be more active than those with lower σz^n.

Figure 13 shows the map of the Montafon range with the distribution of all rock glaciers we kept into the inventory. The red colored polygons describe the shape of identified rock glaciers. The inventory could be filed up with the knowledge about the activities of each rock glacier. Based on the method described above, the rock glaciers were classified into inactive and intact. In the investigation area, there are only a few intact rock glaciers left.

Altogether 193 rock glaciers were detected. From these ten rock glaciers were classified as intact, and 173 as relict. Another ten rock glaciers could not be classified, as

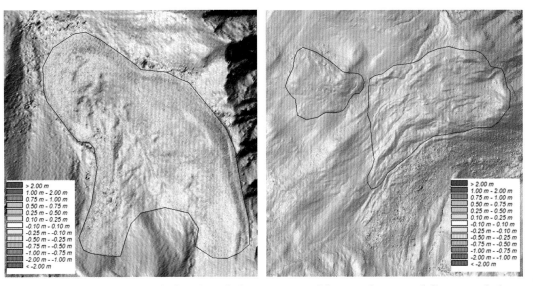

Figure 14: Two examples from the rock glacier inventory of the Montafon Range. (left) intact rock glacier; (right) two relict rock glaciers

they lie out of the area covered by the ALS data set. It was not as easy to distinguish intact rock glaciers seriously between active and inactive without knowledge about the movement rates, as this is the main criterion for a rock glacier being active or inactive (Barsch 1996).

Figure 14 shows two examples of the inventory with their elevation changes from 2004 to 2010. The right image shows an intact rock glacier with an area-averaged surface change of −0.38 m and a σz^n of about 0.41 m. The positive values along the rock glacier front are a visible sign for the movement of that permafrost body. The left image is an example of two relict rock glaciers with nearly no altitudinal changes ($\varDelta z^n$ = −0.03 m and σz^n = 0.02 m).

With the resulting rock glacier inventory, it will be possible to model the overall permafrost distribution in that region. Furthermore it is planned to calculate the horizontal displacement rates of the mapped intact rock glaciers in order to distinguish between inactive or active type.

Combining this dataset with the other existing inventories of Austria (e.g. Tyrol) and using the high resolution multi-temporal ALS data, an accurate large scale permafrost distribution model can be realized. These efforts are planned for subsequent project activities.

4.3   Detecting permafrost evidence on Rofenberg

4.3.1   Data and remote sensing approach

As mentioned in section 2 a test site was installed at the Rofenberg in the Ötztal Alps (Tyrol, Austria) to verify the remote sensing based insights about permafrost induced surface changes. Since 2001, ALS measurements have been carried out regularly at Hintereisferner and the adjacent Rofenberg, resulting in a unique data record of 21 ALS flight campaigns until now. A summary of the available ALS campaigns, and their corresponding system parameters, is given in Sailer et al. (2012). Regarding the accuracy of the ALS data series, Bollmann et al. (2011) indicate an absolute vertical accuracy of 0.07 m with a standard deviation of ± 0.08 m. The results are based on a comparison of dGPS points with the nearest neighbouring ALS points on the relatively flat and smooth glacier tongue of Hintereisferner.

As the experimental design (sensor types, height above ground, geo-referencing and transformation parameters) of all ALS campaigns were nearly uniform, the slope dependent error is very small (± 0.05 m for slope angles < 40°), compared to previously reported values. The delineation of possible permafrost areas, which are characterised by process induced surface altitudinal changes, was based on visual inspections of annual and multi-annual dDTM (differential DTM). In near proximity to the detected areas, stable areas, where no altitudinal changes were observed, are defined. These stable areas are used for i) quantitative correction of the process related altitudinal changes or ii) to discard a specific dataset if necessary (cf. Fig. 15; cf. Sailer et al. 2012).

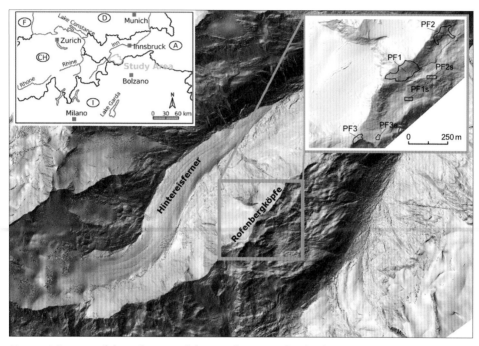

Figure 15: Location of the study area and the exemplary areas of surface lowering (PF I–III). (Cartographic basemap: ALS DTM, Institute of Geography, Innsbruck 2011)

At several small areas near the mountain ridge of the "Rofenbergköpfe" south of the glacier tongue of Hintereisferner, a small but continuous lowering of the surface (cf. Table 7) was detected throughout the whole data series of 22 ALS flights. Those surface changes were assumed to originate from the melting of permafrost or small dead-ice bodies. To verify this assumption, several measurements have been done or prepared.

Table 7: Annual altitudinal changes dz (m) to the reference level 2011 of possible permafrost and stable areas (cf. point-based results in Sailer et al. 2012 and accuracy assessment in Sailer et al. 2013)

			2010	2009	2008	2007	2006	2005	2004	2003	2002	2001
permafrost areas	PF1	dz process	−0.06	0.04	−0.09	−0.33	−0.39	−0.64	−0.36	−0.42	−0.48	−0.96
		dz stable (PF1s)	0.10	0.10	0.10	−0.07	−0.19	−0.45	−0.06	−0.06	−0.15	−0.36
		dz process corrected	−0.16	−0.06	−0.19	−0.26	−0.20	−0.19	−0.30	−0.36	−0.33	−0.60
	PF2	dz process	−0.01	−0.04	−0.03	−0.29	−0.29	−0.69	−0.29	−0.35	−0.40	−0.97
		dz stable (PF2s)	0.11	0.10	0.11	−0.18	−0.18	−0.52	−0.09	−0.02	−0.14	−0.37
		dz process corrected	−0.12	−0.14	−0.14	−0.12	−0.11	−0.17	−0.20	−0.33	−0.26	−0.60
	PF3	dz process	−0.21	−0.21	−0.05	−0.38	−0.32	−0.70	−0.33	−0.29	−0.40	−0.95
		dz stable (PF3s)	−0.06	0.03	0.10	−0.26	−0.17	−0.80	−0.14	0.00	−0.11	−0.42
		dz process corrected	−0.15	−0.24	−0.15	−0.12	−0.15	0.10	−0.19	−0.29	−0.29	−0.53

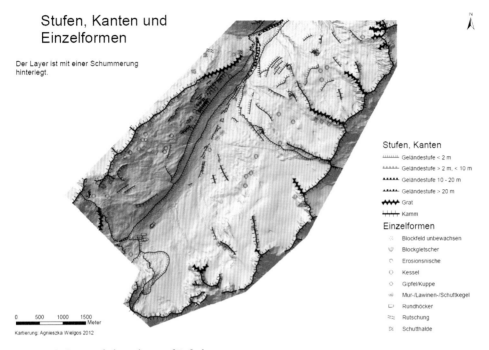

Stufen, Kanten und Einzelformen

Der Layer ist mit einer Schummerung hinterlegt.

Stufen, Kanten
- Geländestufe < 2 m
- Geländestufe > 2 m, < 10 m
- Geländestufe 10 - 20 m
- Geländestufe > 20 m
- Grat
- Kamm

Einzelformen
- Blockfeld unbewachsen
- Blockgletscher
- Erosionsnische
- Kessel
- Gipfel/Kuppe
- Mur-/Lawinen-/Schuttkegel
- Rundhöcker
- Rutschung
- Schutthalde

0   500   1000   1500
Meter

Kartierung: Agnieszka Wielgos 2012

Figure 16: Geomorphological map of Rofenberg

Figure 18 shows the local distribution of the different measurements. The annual surface elevation changes in the detected permafrost areas are very low (between –0.05 m and –0.10 m per year). Therefore, the significance of the process dependant measured topographic change rates was assessed with special regard to the accuracy of the ALS data, the magnitude of the process, the time lapse between the single ALS-campaigns and frequency and disturbing factors (e. g. snow cover). Results gained in significance with increasing time laps between the ALS-campaigns, the frequency of flight campaigns and if disturbing factors (e. g. snow cover) can be excluded.

4.3.2   Methods

4.3.2.1   Geomorphologic mapping
In the course of geomorphologic mapping on Rofenberg different approaches have been applied. Firstly, intensive field investigations have been conducted. In a second step, the mapped surface structures have been cross checked with a geoinformatic approach. Therefore the ALS data have been processed with an automatic breakline detection tool implemented in GRASS GIS (Rutzinger et al. 2010). To get clear results the mapped and automatic derived surface structures have been adjusted using an orthoimage and a shaded relief (acquisition time September 2010) and a difference layer of the DTMs of 2006 and 2010. The geomorphological mapping has

been conducted using a legend for high mountain geomorphology, which is based on GMK 25 (cf. Kneisel et al. 1998; Fig. 16).

4.3.2.2 BTS measurements and BTS loggers

Measurements of ground temperatures of the winter snow cover (the so called BTS method) are a well-established technique to map mountain permafrost. Hereby, a probe is pushed through the snow cover to the ground surface. An important boundary condition for the successful application of this method is a sufficiently thick (at least 60–80 cm) snow cover. The snow cover, with its low thermal conductivity, insulates the ground from short-term fluctuations in air temperature. If the winter snow

Table 8: Coordinates and terrain description of the 15 temperature loggers displaced on Rofenberg

	UTM-Coordinates	Z [m a. s. l]	Description of area around logger
1	637163 E, 5184507 N	3,033	flat crest, stone diameter: 2–20 cm Soil and fine-grained substrate
2	637123 E, 5184101 N	3,041	flat area, flat crest, fine-grained material (gravel / sand), stone diameter 5–20 cm (10% > 20 cm), no local shading, shallow weathered layer on top of bedrock (ca. 0.5 m)
3	637057 E, 5184143 N	3,048	logger underneath small boulder (diameter 50 cm), east-facing, flat, small crest surrounded by surface water, stones 10–20 cm, no local shading, snow rich area
4	636963 E, 5184175 N	3,070	logger hanging on a lash (40 cm long) in blocky area, always in shade, no fine-grained material, boulder: 0.5–1 m, crest, rock fall material
5	636310 E, 5184489 N	3,203	boulder material d = 60 cm, no fine-grained material, SE, 23°
6	636876 E, 5184543 N	3,213	SE, 25°, boulder with d = 40 cm
7	636950 E, 5184586 N	3,173	East, 35°, boulder with d = 50 cm
8	637030 E, 5184582 N	3,164	blocky material d = 60 cm, individual "moss pillows", east, flat area, 5°
9	637058 E, 5184592 N	3,144	flat, depression with d = 30 m, sand / gravel, stones d = 20 cm
10	637088 E, 5184688 N	3,132	small basin with d = 20 m, depression, snow rich, flat, no topographic shading
11	637146 E, 5184743 N	3,125	flat, crest, no topographic shading, stones with d = 20 cm, some fine-grained material, sporadic moss/grass
12	637066 E, 5184779 N	3,156	SEE, 18°, „flat area" with 80 m x 30 m, frost affected surface, sandy, stones < 10 cm individual with 30 cm, distance to main crest ca. 20 m horizontal
13	637167 E, 5184777 N	3,112	depression, snow rich, logger hanging on a lash 10 cm underneath surface, blocky, depression NE facing, 25°
14	637517 E, 5184863 N	3,016	same exposition as logger 15 (SO), depression (below crest 50 vertical m, 150 horizontal m), fine-grained material available, logger on "soil" but covered with flat stone, stone diameter < 40 cm
15	637707 E, 5184684 N	2,941	flat area, grass, fine-grained material (sand, gravel), few stones, micro exposition South,

Figure 17: BTS measurements on Rofenberg, March 2011, photograph by L. Rieg

cover is sufficiently thick and surface melting is still negligible in mid- to late winter, the BTS values remain nearly constant and are mainly controlled by the heat transfer from the upper ground layers, which in turn is strongly influenced by the presence or absence of permafrost. Values obtained over permafrost are below –3 °C. Values between –2 and –3 °C represent the uncertainty range of the method and/or marginally active permafrost with thick active layer, which does not totally refreeze during winter. At permafrost-free sites or sites with inactive permafrost the measured values are above –2 °C (Haeberli 1996).

In addition to spot measurements of the BTS, 15 temperature loggers (see Table 8) were placed to record the BTS and year-round near-surface temperatures. From the latter data the mean annual ground surface temperature (MAGST) can be calculated. According to van Everdingen (1998), permafrost exists if the MAGST is perennially below 0 °C. Concerning the MAGST, the thermal offset, which is defined as the difference between mean annual temperature at the permafrost table and the mean annual ground surface temperature, has to be taken into account. Burn and Smith (1988) have suggested that permafrost may be in equilibrium, or aggrading, even under conditions where the mean annual ground surface temperature is slightly above 0 °C.

Figure 18: Temperature logger distribution. Observation point 14 located at Rofenberg at 3,000 m a. s. l., photograph by E. Bollmann

In spring 2011 (30.03.2011) and 2012 (2.04.2012) the base

temperature of the snow pack (BTS) was measured in the whole area close to the end of the winter snow accumulation period, to get a general idea of the possible distribution of permafrost or ice in the underground. Five probes were placed in a circular arrangement as it is shown in Figure 17. The data were collected between altitudes above and beneath the lower boundary of discontinuous permafrost.

For further investigations on the occurrence and distribution of permafrost, 15 BTS loggers were installed in order to measure the temperature during the winter. This allows recording the potential freezing and thawing of the underlying soil. The loggers are distributed over different elevation levels to detect the lower boundary of discontinuous permafrost (Fig. 18 and Table 8).

4.3.2.3 *Geophysical surveys*

Geophysical methods are particularly suitable for geomorphological investigations, since the knowledge of structure, layering and composition of the subsurface at different scales are key parameters for geomorphological problems. Georadar, geoelectric and seismic methods were used to detect permafrost, with each geophysical method being applied on all profile lines. This parallel application enabled us to compare and cross-validate the results of the three techniques. After the analyses of the single datasets, a tomography including all results was created.

In July and September 2011, geophysical measurements were carried out at two selected sites on Rofenberg (Fig. 19). Three profiles (B, C, D) were measured at a height of about 3,200 m. Two of them were parallel and one was crossing them both.

Figure 19: Geophysical field measurements on Rofenberg, July 2011, photograph by C. Klug

Each profile had a length of about 100 m. The spacing of the electrodes was 2 m for geoelectrical measurements using Wenner geometries. The geophone spacing for seismic measurements was about 4 m and a shot spacing of 4 m was applied. For georadar measurements 50 MHz, 100 MHz and 200 MHz antennas were used. Moreover, another profile (A) was measured at a height of around 2,900 m a.s.l. with nearly the same calibrations (cf. Fig. 20).

4.3.3   Results and Interpretation

The occurrence of settlement features that might be related to melting permafrost provides a first geomorphological evidence of permafrost occurrence. This verifies the assumption that the small but continuous lowering of the surface (cf. Table 7), which was detected throughout the whole data series of 22 ALS flights, originates from the melting of permafrost.

According to BTS measurements, performed in spring 2011 and 2012, the higher parts of the investigated area between 3,200 and 3,300 m a.s.l., are underlain by permafrost. Measurements were recorded that are mostly below –3 °C, moreover at three measurements are at least within the uncertainty range of the method (–2 to –3 °C). In the lower parts of Rofenberg, permafrost is mapped as improbable (Fig. 20).

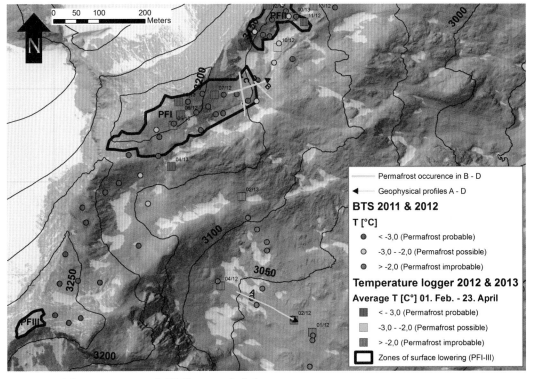

Figure 20: BTS measurements and GST lgging on Rofenberg

These results are supported by findings from the geophysical measurements and measured year-round near-surface ground temperatures for the period September 2011 until September 2012, which were recorded using fifteen miniature loggers.

In order to assess ground thermal conditions, the temperature loggers were placed in the upper parts of the supposed permafrost areas and near the geophysical profiles. Twelve of the 15 loggers were obviously covered by a thick enough snow cover, as there are no high frequency variations visible in the graph and the values decrease gradually until the final more or less constant temperature is reached in the mid to late winter months. Thus, the bottom temperatures of the snow cover can be analysed according to the three distinguished BTS classes. From the beginning of May an increase of the temperatures to 0 °C is visible; in the following weeks the temperatures remain constant around 0 °C (zero curtain). This feature is typical for mountain permafrost with dry freezing in autumn without development of a zero curtain and wet thawing in spring with pronounced zero curtain (cf. Kneisel & Kääb 2007).

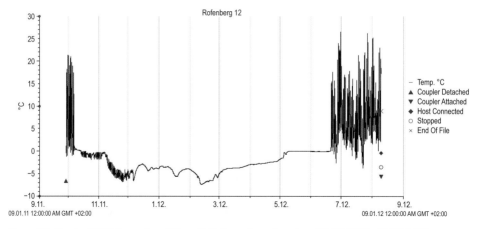

Figure 21: Result of BTS-logger (no. 12) on Rofenberg. Recording time: 27.09.2011–12.09.2012

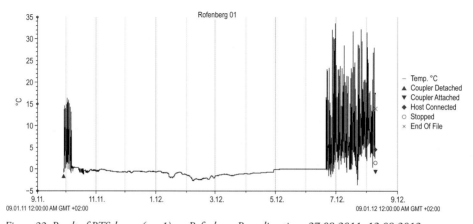

Figure 22: Result of BTS-logger (no. 1) on Rofenberg. Recording time: 27.09.2011–12.09.2012

In the winter period the loggers 6 to 12 recorded BTS values that indicate permafrost. Mean annual ground surface temperatures could be calculated for the period 27 September 2011 to 12 September 2012. The results with negative mean annual ground temperatures (e. g. logger 12: –1.43 °C) point to permafrost favourable conditions during the measurement period. The survey presented in Figure 21 was performed on the uppermost part of Rofenberg, where BTS measurements point to the presence of permafrost (Fig. 20). In contrast the logger 1 (Fig. 22), situated in the lowest part of Rofenberg, shows positive MAGST, which is in good accordance with the BTS measurements there and indicates that permafrost is improbable.

4.3.3.1 Ground Penetrating Radar (GPR) results

GPR results for the upper three profiles (cf. Fig. 23, 24 and 25) show reflectors close to the surface that are all present for the most part of the profiles. Deeper regions of the profiles show continuous reflectors in a depth of about 10 and 22 m for profile D (Fig. 23) and profile B (Fig. 25). Profile C shows a close to the surface reflector that follows the whole profile as well (Fig. 24). Deeper regions in this profile cannot be interpreted by using radar due to low signal to noise ratio. Similar depths of reflectors for those three profiles are consequentially because all profiles are based in the same area. Profile A (Fig. 26) being based several hundred meters away shows reflectors in a depth of about three and nine meters, but they are less significant than in the upper zone.

4.3.3.2 Electric Resistivity Tomography (ERT) results

The ERT results for all profiles show resistivities later being interpreted as permafrost. Table 9 gives the specific resistivities of relevant materials. Especially profile C (Fig. 27) and B (Fig. 28) fit together well. You can see a low resistivity zone (ca.

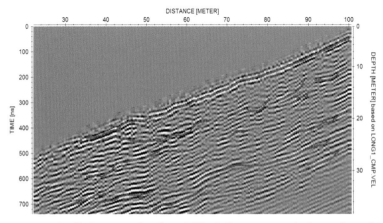

Figure 23: Profile D measured with 50 MHz. Reflector 1 proceeds close to the surface in a depth of about 2 m. Reflector 2 starts at a depth of 2 m, descending until 10 m depth after short distance being visible until the end of the profile. Deeper regions show further reflectors in a depth of about 22 m.

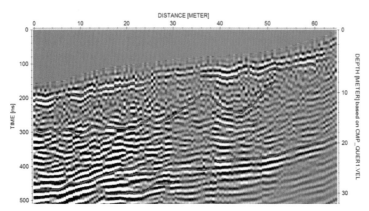

Figure 24: Profile B measured with 50 MHz. This profile shows quite similar results as profile D. Reflector 1 proceeds in a depth of about 2 to 3 m and is visible until the end of the profile. Another reflector is visible in a depth of about 10 m. The reflections in about 22 m depth are visible in this profile as well

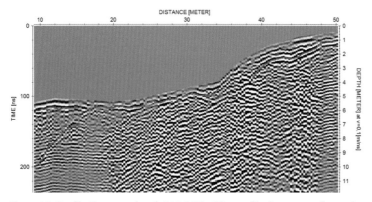

Figure 25: Profile C measured with 200 MHz. This profile shows one reflector close to the surface in a depth of about 2 m. Further reflectors in deeper regions can't be interpreted due to low signal to noise ratio

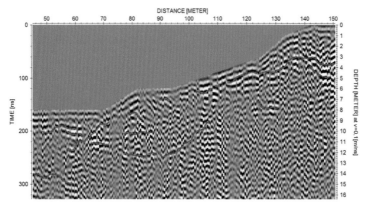

Figure 26: Profile A measured with 100 MHz. One reflector in a depth of about 2 m can be seen in this profile. Another reflector is interpreted in a depth of about 8 m

1,500 Ωm) starting at 32 m x-distance up
to 3 m depth which fit to sand and gravel
of the surface. This area is followed by a
zone up to 10 m depth with high resisitivi-
ties which fit well to permafrost. Profile B
shows a high resistivity area starting from
28 m x-distance up to a depth of 10 m.
This also fits to permafrost. The low resistivity zone (ca. 700 Ωm) at the beginning
of profile B is related to a small lake to the
profile so this area is saturated with water.
In profile A no significant resistivity zone
concerning the permafrost was found.

Table 9: Specific resistivities of relevant materials (Knödel et al. 1997; Maurer & Hauck 2007)

Materials	Range of resistivity [Ωm]
Air	infinite
(Ground-) Water	10–300
Sand/Gravel	$100–10^4$
Metamorphic rock	$1\,000–10^5$
Permafrost	$5\,000–10^6$
Glacial ice	$10^5–10^7$

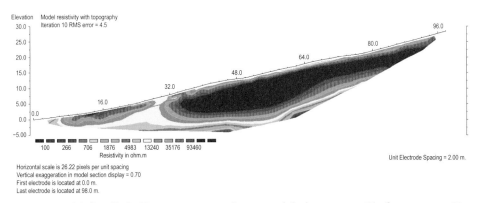

Figure 27: ERT of profile D. You can see two purple areas with high resistivities. The first area up to 28 m x-distance fit to the ice rich gravel on the ground. The area starting at 32 m x-distance is interpreted as per-mafrost. The upper 2 m have lower resistivities fitting to the surface gravel

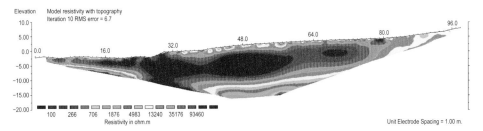

Figure 28: ERT of profile C. The purple red area starting at 32 m x-distance fits to permafrost. The area from zero to 32 m x-distance was covered by snow and there was a dyke between 20 m and 32 m filled with snow, therefore in this area you can see big resistivities. Up to 3 m depth starting at 32 m x-distance the resisitvities fit to the surface gravel

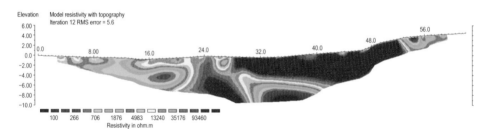

Figure 29: ERT of Profile B. Here you can see a blue are between 8 to 20 m x-distance which is related to a sea near to the profile and therefore the underground is saturated with water. The big purple area starting at 30 m x-distance fits to permafrost

4.3.3.3 Seismic Refraction Tomography (SRT) results

It was only possible to measure the SRT at profile D and 70% of profile C due to technical problems. Both profiles show slow seismic velocities up to a depth of 2 to 3 m (Fig. 30 and 31; ca. 1,000 m/s) followed by an area of velocities between 2,500 and 3,500 m/s (cf. Table 10). The first area represents the surface with sand and gravel and the deeper area fits to the velocity range of Permafrost. In Figure 30 you can also spot a zone with velocities up to 5,000 m/s which fits to metamorphic rock and could represent the bedrock.

For evaluating an underground model of the measured area all three methods

Table 10: Seismic velocities for relevant materials (Hecht 2001; Knödel et al. 1997; Maurer & Hauck 2007)

Material	Seismic Velocity (m/s)
Air	330
Water	1,400–1,600
Sand/Gravel	300–2,000
Boulder	600–2,500
Morainal material / Glacial sediments	1,500–2,700
Metamorphic rock	3,000–5,700
Permafrost	2,400–4,300
Glacial ice	3,100–4,500

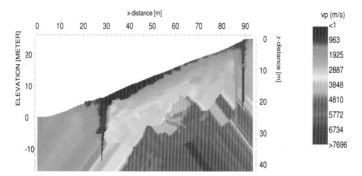

Figure 30: SRT Velocity model of Profile D. You can see the sand/gravel area (blue) up to 3 m depth. The green/yellow area starting at 25 m x-distance fits to Permafrost. Up to 25 m x-distance in depths up to 3 m the velocities are higher than in the other surface areas due to visible ice rich gravel. In depths starting at 15 m depth the red area fits to metamorphic rock with velocities up to 5,000 m/s

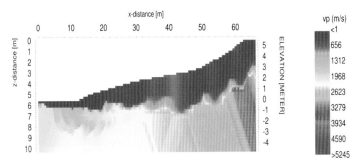

Figure 31: SRT Velocity model of Profile B. You can spot the blue area of gravel like velocities in depths up to 3 m following by an area (yellow/orange) which fits to permafrost with velocities over 2,300 m/s. Due to technical problems it was only possible to measure 70% of this profile. 65 m x-distance is equivalent to 96 m x-distance of the ERT

were combined by overlaying the results. The finished models are shown in Figure 32–34. Permafrost was interpreted in all four profiles.

Figure 32 shows the model of profile D where you can see a permafrost layer with 64 m length and up to 10 m depth. The permafrost layer of profile B (Fig. 33) has

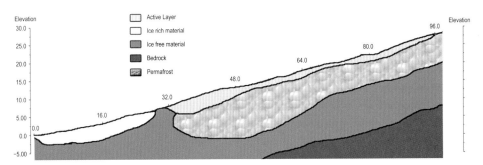

Figure 32: Underground model of profile D. The active layer could be sawn clearly and permafrost is also interpreted. Furthermore the ice rich surface was represented in the data

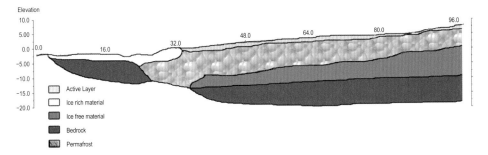

Figure 33: Underground model of Profile B. The snow dyke is detected in the data and also a permafrost body could be interpreted as well as the active layer

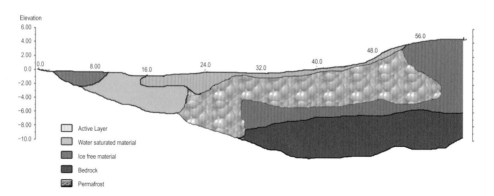

Figure 34: Underground model of profile D. A clear permafrost layer could be interpreted. The active layer is also represented in the data as well as a water saturated area as result of the sea near the profile

a length of 75 m and a mean depth of 5 m. In Figure 34 you see the model of profile C with a permafrost layer of about 34 m length and 2–6 m depth. In the area of profile A no permafrost was detected. The proportions are about a length of 60 m and a depth of 13 m.

5   Overall Conclusions

Several geomorphologic changes due to melting permafrost were detected including the occurrence of rock glaciers. The studies of Sailer et al. (2012), which are related to the project permAfrost, provide detailed investigations of the quantification of geomorphodynamics. Surface elevation changes or single rock fall events, which may be caused by thawing permafrost, were detected with the help of multi-temporal ALS data.

All presented methods and workflows have to be applied in combination in order to provide an operational methodology for on-going, spatially continuous and area-wide monitoring of geomorphological significant permafrost change. Especially multi-temporal ALS is useful for the large scale investigations. As a second step, the findings of that remote sensing based approach have to be verified with in-situ investigations like BTS and / or geophysical measurements (e. g. ERT). The results we described above demonstrate the capability detecting permafrost spatially continuous and area-wide. However, successful investigations depend strongly on the data quality (e. g. ALS). Even financial efforts, especially when acquiring ALS flight campaigns, may not be underestimated, as this technology is still very cost-intensive.

Geomorphological observations, ALS based remote sensing approaches, BTS measurements and geophysical measurements on the Rofenberg were used to obtain an integrative analysis of a highly complex periglacial landform. A valuable aspect of modern geophysical techniques and their application in geomorphology is the ability to image the subsurface.

Table 11: Overview of the applied methods

Method	Pros and cons	Ability	Requirements
ALS	**+** • Provides operational tool for spatially continuous and area-wide monitoring of permafrost (PF) change • High resolution (0.5 m) possible • High vertical accuracy • Operational stage • Contact free **−** Cost intensive Weather condition dependent	• High vertical accuracy allows detection of PF degradation • Possibility of individual flight planning (research area, resolution, acquisition date)	• Flyable weather • High temporal resolution (annual, inter-annual, decadal) for different processes • Data storage, management and processing tools • Sufficient point density
Aerial images	**+** • Provides an operational tool for spatially continuous and area-wide monitoring of PF change • Long-time monitoring (back to 1950s) • High horizontal accuracy • Operational stage • Combinable with ALS • Contact free **−** • Less cost intensive • Shadowing and bad contrast effects • Lower vertical accuracy and lower resolution than ALS	• Long-time monitoring (back to 1950s) • High horizontal accuracy allows mapping of morphological PF features	• Avoiding of shadowing and bad contrast effects • High temporal resolution (annual, decadal) for different processes • Overlapping images for DTM generation
Geophysics	**+** • Tool for local monitoring of permafrost • Subsurface surveys **−** • Less information about surface changes • Not area-wide • Time consuming • Bulky instruments for field investigation	• Provision of subsurface information • Information of subsurface PF distribution • Ice content estimation of the ice body	• Combination of geophysical methods for better results • Accessibility • Expert know-how
BTS	**+** • Tool for local monitoring of permafrost • Good significance on PF occurrence **−** • Time consuming • Not area-wide	• In situ evidence of PF presence • Verification of PF distribution models	• Snow conditions (> 60 cm snow cover) • Consistent spatial distribution of point measurements • Measurement at the end of accumulation period (BTS-probing)
Geomorphological mapping	**+** • Good first basis for PF investigation • Less information about surface changes **−** • Time consuming • Subjectivity of the results	• Ability for first research questions • Detection of geomorphologic features to gain information about recent or past PF activities	• Combination of field investigations and remote sensing necessary

The agreement between the ALS based measurements and geophysical surveys presented on Rofenberg demonstrated that the investigation of surface changes by DTM differencing in cold mountain environments can be a suitable tool to detect surface deformation as a proxy for ice-containing permafrost. The combined approach allows a better understanding of the recent interplay between ice content and surface deformation. Hereby, integrative analyses have the potential to improve the understanding of permafrost-related periglacial landforms, which may exhibit active (supersaturated with ice and creeping), inactive (degrading ice and almost no creeping) and relict (ice-free, no creeping) sediment bodies in close proximity.

Concerning the main objectives, namely the detection of locations, where melting of permafrost leads to geomorphologic changes, and the provision of an operational methodology for on-going, spatially continuous, area-wide monitoring of geomorphological significant permafrost change, we tried out different methods. Table 11 gives an overview of the applied methods, their pros and cons and their ability in permafrost research and their requirements for good and interpretative results concerning permafrost degradation.

6 References

Barsch, D. 1996: *Rock glaciers. Indicators for the Present and Former Geoecology in High Mountain Environments.* Berlin.

Baltsavias, E. 1999: Airborne laser scanning: basic relations and formulas. *ISPRS Journal of photogrammetry and remote sensing* 54: 199–214.

Baltsavias, E.P., E. Favey, A. Bauder, H. Boesch & M. Pateraki 2001: Digital surface modelling by airborne laser scanning and digital photogrammetry for glacier monitoring. *Photogrammetric Record* 17, 98: 243–273.

Bauer, A., G. Paar & V. Kaufmann 2003: Terrestrial laser scanning for rock glacier monitoring. In: Phillips M., S.M. Springman & L.U. Arenson (eds.): *Permafrost: Proceedings of the 8th International Conference on Permafrost*, Zurich, Switzerland, 21-25 July 2003: 55–60.

Bollmann, E., R. Sailer, C. Briese, J. Stötter, & P. Fritzmann 2010: Potential of airborne laser scanning for geomorphologic feature and process detection and quantification in high alpine mountains. *Zeitschrift für Geomorphologie* 54 (Suppl. Issue 2): 83–104.

Bollmann, E., J. Abermann, C. Klug, R. Sailer & J. Stötter (2012): Quantifying Rock glacier Creep using Airborne Laserscanning. A case study from two Rock glaciers in the Austrian Alps. In: Hinkel, K.M. (ed.): *Proceedings of the Tenth International Conference on Permafrost*, Salekhard, Russia, 25-29 June 2012. Vol. 1: International Contributions: 49–54.

Dowdeswell, J.A. & T.J. Benham 2003: A surge of Perseibreen, Svalbard, examined using aerial photography and ASTER high resolution satellite imagery. *Polar research*, 22,2: 373–383.

Haeberli, W. 1982: Creep of mountain permafrost: Internal structure and flow of alpine rock glaciers. *Mitteilungen der Versuchsanstalt für Wasserbau, Hydrologie und Glaziologie ETH Zürich* 77: 1–142.

Haeberli, W. & G. Patzelt 1982: Permafrostkartierung im Gebiet der Hochebenkar-Blockgletscher. Obergurgl, Ötztaler Alpen. *Zeitschrift für Gletscherkunde und Glazialgeologie* 18, 2: 127–150.

Haeberli, W., B. Hallet, L. Arenson, R. Elconin, O. Humlum, A. Kääb, V. Kaufmann, B. Ladanyi, N. Matsuoka & D. Vonder Mühll 2006: Permafrost creep and rock glacier dynamics. *Permafrost and periglacial processes* 17, 3, 189–214.

Hausmann, H., K. Krainer, E. Brückl & W. Mostler 2007: Internal Structure and Ice Content of Rei-chenkar Rock Glacier (Stubai Alps, Austria) Assessed by Geophysical Investigations. *Permafrost and Periglacial Processes* 18: 351–367.

Hecht, S. 2001: *Anwendung refraktionsseismischer Methoden zur Erkundung des oberflächennahen Unter-grundes.* Stuttgarter Geographische Studien, 131. Stuttgart.

Hodgsonn, M. & P. Bresnahan 2004: Accuracy of airborne Lidar-derived elevation: Empirical assess-ment and error budget. *Photogrammetric engineering and remote sensing* 70, 3: 331–339.

Hulbe, C.L., T.A. Scambos, T. Youngberg & A.K. Lamb 2008: Patterns of glacier response to disinte-gration of the Larsen B ice shelf, Antarctic Peninsula. *Global and planetary change* 63: 1–8.

Jackson, M., I.A. Brown & H. Elvehøy 2005: Velocity measurements on Engabreen, Norway. *Annals of glaciology* 42: 29–34.

Jokinen, O. & T. Geist 2010: Accuracy aspects in topographical change detection of glacier surface. In: Pellikka, P. & W.G. Rees (eds.): *Remote sensing of glaciers. Techniques for topographic, spatial and the-matic mapping of glaciers.* London: 269–283.

Kääb, A. & M. Vollmer 2000: Surface geometry, thickness changes and flowfields on creeping moun-tain permafrost: automatic extraction by digital image analysis. *Permafrost and Periglacial Processes* 11, 4): 315–326.

Kääb, A. 2004: *Mountain glaciers and permafrost creep. Research perspectives from earth observation tech-nologies and geoinformatics.* Habilitation. Department of Geography, ETH Zürich.

Kääb, A. 2010: Aerial photogrammetry in glacier studies. In: Pellikka, P. & W.G. Rees (eds.): *Remote Sensing of Glaciers. Techniques for Topographic, Spatial and Thematic Mapping of Glaciers.* London: 115-136.

Kääb, A., R. Frauenfelder & I. Roer 2007: On the reaction of rock glacier creep to surface temperature variations. *Global and planetary change* 56: 172–187.

Kääb, A., W. Haeberli & H. Gudmundsson 1997: Analysing the creep of mountain permafrost using high precision aerial photogrammetry: 25 years of monitoring Gruben rock glacier, Swiss Alps. *Per-mafrost and periglacial processes* 8: 409–426.

Kääb, A., V. Kaufmann, R. Ladstädter & T. Eiken 2003: Rock glacier dynamics: implications from high-resolution measurements of surface velocity fields. In: Phillips M., S.M. Springman & L.U. Arenson (eds.): *Permafrost: Proceedings of the 8th International Conference on Permafrost*, Zurich, Switzerland, 21-25 July 2003: 501–506.

Kääb, A., K. Isaksen, T. Eiken & H. Farbot 2002: Geometry and dynamics of two lobe-shaped rock glaciers in the permafrost of Svalbard. *Norsk geografisk tidsskrift* 56: 152–160.

Kaufmann, V. & R. Ladstädter (2002): Monitoring of active rock glaciers by means of digital photo-grammetry. In: *Proceedings of the ISPRS Commission III Symposium "Photogrammetric Computer Vi-sion"*, Graz, Austria, 9-13 September 2002. IAPRS 34, 3B: 108–111.

Kaufmann, V. & Ladstädter, R., 2003. Quantitative analysis of rock glacier creep by means of digital photogrammetry using multi-temporal aerial photographs: two case studies in the Austrian Alps. In: Phillips M., S.M. Springman & L.U. Arenson (eds.): *Permafrost: Proceedings of the 8th International Conference on Permafrost*, Zurich, Switzerland, 21-25 July 2003: 525–530.

Knödel, K., H. Krummel & G. Lange 1997: *Handbuch zur Erkundung des Untergrundes von Deponien und Altlasten. Band 3. Geophysik.* Bundesanstalt für Geowissenschaften und Rohstoffe. Berlin.

Krainer, K. & W. Mostler 2000: Reichenkar Rock Glacier: a Glacier-Derived Debris-Ice System in the Western Stubai Alps, Austria. *Permafrost and Periglacial Processes* 11: 267–275.

Krainer, K. & W. Mostler 2002: Hydrology of active rock glaciers in the Austrian Alps. *Arctic, Antarctic and Alpine Research* 34: 142–149.

Krainer, K. & W. Mostler 2006: Flow velocities of active rock glaciers in the Austrian Alps. *Geografiska Annaler* 88 (a): 267–280.

Krainer, K., W. Mostler & N. Span 2002: A glacier-derived, ice-cored rock glacier in the western Stubai Alps (Austria): Evidence from ice exposures and ground penetrating radar investigation. *Zeitschrift für Gletscherkunde und Glazialgeologie* 38: 21–34.

Kraus, K. 2004: *Photogrammetrie, Band 1. Geometrische Informationen aus Photographien und Laserscanneraufnahmen.* Berlin.

Lambiel, C. & R. Delaloye 2004: Contribution of real-time kinematic GPS in the study of creeping mountain permafrost: examples from the Western Swiss Alps. *Permafrost and Periglacial Processes* 15: 229–241.

Maurer, H. & C. Hauck 2007: Instruments and Methods – Geophysical imaging of alpine rock glaciers. *Journal of Glaciology* 53, 180: 110–120.

Pillewizer, W., 1957. Untersuchungen an Blockströmen der Ötztaler Alpen. In: Fels E., H. Overbeck & J.-H. Schultze (eds.): *Geomorphologische Abhandlungen: Otto Maull zum 70. Geburtstage gewidmet.* Abhandlungen des Geographischen Institutes der Freie Universität Berlin 5. Berlin: 37–50.

Rignot, E., B. Haller & A. Fountain 2002: Rock glacier surface motion in Beacon Valley, Antarctica, from synthetic-aperture radar interferometry. *Geophysical research letters* 29, 12: 48-1–48-4. doi:10.1029/2001GL013494.

Roer, I. 2005: *Rock glacier kinematics in a high mountain geosystem.* Dissertation. Department of Geography, University of Bonn.

Roer, I., W. Haeberli, M. Avian, M. Kaufmann, R. Delaloye, C. Lambiel & A. Kääb 2008: Observations and considerations on destabilizing active rock glaciers in the European Alps. In: Kane, D.L. & K.M. Hinkel (eds.): *Proceedings of the Ninth International Conference on Permafrost,* Fairbanks, Alaska, 29 June-3 July 2008. Vol. 2: 1505–1510.

Scambos, T.A., G. Kvaran & M.A. Fahnestock 1999: Improving AVHRR resolution trough data cumulation for mapping polar ice sheets. *Remote sensing of environment* 69: 56–66.

Scambos, T.A., M.J. Dutkiewitcz, J.C. Wilson & R.A. Bindschadler 1992: Application of image cross-correlation software to the measurement of glacier velocity using satellite data. *Remote sensing of environment* 42: 177–186.

Schneider, B. & H. Schneider 2001: Zur 60jährigen Messreihe der kurzfristigen Geschwindigkeitsschwankungen am Blockgletscher im Äußeren Hochebenkar. *Zeitschrift für Gletscherkunde und Glazialgeologie* 37, 1: 1–33.

Stearns, L. & G. Hamiltom 2005: A new velocity map for Byrd Glacier, East Antarctica, from sequential ASTER satellite imagery. *Annals of glaciology* 41: 71–76.

Strozzi, T., A. Kääb & R. Frauenfelder 2004: Detecting and quantifying mountain permafrost creep from in-situ, airborne and spaceborn remote sensing methods. *International journal of remote sensing* 25, 15: 2919–2931.

Vietoris, L. 1958: Der Blockgletscher des Äußeren Hochebenkares. *Gurgler Berichte* 1: 41–45.

Vietoris, L. 1972: Über die Blockgletscher des Äußeren Hochebenkars. *Zeitschrift für Gletscherkunde und Glazialgologie* 8: 169–188.

Vitek, J.D. & J.R. Giardino 1987: Rock glaciers: a review of the knowledge base. In: Giardino, J.R., J.F. Shroder Jr. & J.D. Vitek (eds.): *Rock Glaciers.* London: 1–6.

Wehr, A. & U. Lohr 1999: Airborne laser scanning – an introduction and overview. *ISPRS Journal of photogrammetry and remote sensing* 54: 68–82.

Whalley, W.B. & H.E. Martin 1992: Rock glaciers: II models and mechanics. *Progress in physical geography* 16: 127–186.